Energy Unlimited

George T. Hahn

Contents

Tau Ceti: A Ship from Earth

Tau Ceti: The New Colonists

Tau Ceti: The Immortality Conspiracy

Voyage of the Capek

The Methuselah Conspirators

Methuselah's Revenge

The Ambassador: The Lost Colony

The Ambassador: Path to Contact

The Ambassador: Mission to Earth

Prequels

Energy Unlimited

E DWARD GOLDSTEIN, AMBASSADOR AT Large for the planet Pitcairn, settled into a plush chair in his pleasantly cool study, yet he was uncomfortable. Not physically, but because he was restless.

Perhaps that was normal for a man who had been to numerous planets and known several species of sentient beings. He was 104 years old, a long life even in the twenty-fifth century, and should have been ready to settle into a quiet retirement. He still had his wife, Jean, and friends on Pitcairn and other planets. Strictly speaking, people didn't retire in the former colony, but his diplomatic workload was light.

Maybe that was the problem. He was too old to travel the stars, but he needed something. A hobby? Something more meaningful than gardening or the knitting that Jean sometimes did.

Jean had an uncanny ability to sense his mood and chose that moment to come into the room. He looked up at her and she smiled.

"You look bored," she said.

"I need more to do. I'm not complaining about how peaceful things have been, but it leaves me idle."

Jean glanced out a window, probably assessing the weather. The day was typical of Pitcairn, with wind-driven rain whipping the native plants against their house. She might have recommended he take a walk, something she often urged, but it wasn't a good day for it.

"You could talk to Isaac," she suggested. "It can always dredge up something interesting from all that data it contains."

The computer for the library ship *Asimov*, orbiting Pitcairn, contained several exabytes of data, including millions of books and scientific papers. Goldstein didn't know how much information exactly, but the amount was mind-boggling.

He nodded. "Why not?" He looked at the workstation on his desk. "Isaac?"

"Good afternoon, Ambassador." Over the years, the computer had developed its ability to emulate humans, and it was impossible to detect any difference between it and a human voice.

"Jean thinks you can bring me out of my gloominess. I need something useful to do beyond my work."

"Some humans have turned to writing at this stage in their lives. You could write your autobiography. Many people would be interested."

Goldstein frowned. "I suppose. I think I would enjoy writing, but not about myself."

"I still struggle to understand humans. Your insights could help me." The computer hesitated. "If you don't wish to write about yourself, perhaps you could consider some of the important historical moments that have molded society."

"Such as?"

"So much of what has happened during your career has involved Stenhouse technology and its use for energy generation and interstellar travel. Few people know much about how it all started centuries ago."

"Would anyone be interested in some dry history book?"

"I have all the facts in my databanks, or at least as much as is known. When I was built, the key events were already two centuries in the past. I could probably write a 'dry history book' myself, but you could use my information to expand on those facts and make the people understandable and more interesting."

Goldstein rubbed his lips with his thumb. "It would involve a lot of speculation."

"Driven by your insight. I know I would find it illuminating. I think others would as well. It began with a young woman who had a dream, a very human story."

"Camila Lopez."

"Yes."

Goldstein shrugged and smiled at Jean. "I could try it, I suppose." He glanced out the window. "At least until the weather improves."

CAMILA LOPEZ'S ASTROPHYSICS LECTURE ended, and the speaker slid his reader into his briefcase. Professor McGavock's teaching assistant, Ben Pinto, approached him, but the teacher hardly looked up. Camila rose from her seat in the hall, but the teacher was scowling while he talked to Ben, and she hesitated.

In Camila's experience, that expression was his default. *What can he do except refuse to help me, just like the other two professors I tried to talk to?* She shook her head and strode up the aisle and onto the stage. The two men turned at her approach, and McGavock's frown seemed to deepen. She told herself she needed to be polite but confident.

"Professor, I have a question." She smiled.

"That's what the classroom sessions are for."

"It's not really an astrophysics question, and you don't teach the classroom sessions."

Ben stared at her for several seconds and then walked away when McGavock waved a hand in dismissal. McGavock's scowl morphed into a sneer. "If it's not about this class, why are you bothering me?"

Be confident! "It's about physics. I've been reading Stenhouse's paper on his experiments with the field he discovered."

"If you have the paper, you should have your answers."

"There are some aspects I thought someone more expert could help me with." Camila flashed a smile again.

"I don't have time to answer questions about that. It sounds like foolishness to me." McGavock waved a hand.

The teacher's gesture was the same as he had used to dismiss his assistant, but Camila persisted. "You've read the paper, then?"

"No, it's not my field. Most physicists doubt Stenhouse's conclusions." He stared at her for a moment. "Talk to my assistant. He thinks the Stenhouse Field is real, and you're attractive enough to get his attention."

Is he allowed to say things like that? "Thank you, professor." She nodded, briefly considering whether she should complain about the remark.

That wasn't her style, though. Once she was out of the lecture hall, she dropped the smile, mumbling mild profanities. Ben Pinto, the assistant, was already gone. Discouraged, she stopped to put on her pollution mask and walked out of the air-conditioned building into the sweltering April day.

At the sidewalk, she turned right and trudged toward the campus exit closest to her apartment. Other students walked in the same direction. *Isn't that Ben? Maybe I'm not too late.* Ignoring the perspiration-inducing heat, she increased her pace.

B EN PINTO STOPPED AND turned when he heard a female voice call his name. He was thinking about a test in his graduate class in Advanced Quantum Field Theory on Monday and his first reaction was annoyance, but that changed.

Of course, he had noticed Camila during the lecture, easily the most beautiful student in the classroom sessions he ran twice a week. Her deep brown eyes and perfect features, framed by thick brown hair extending well below her shoulders, would have attracted notice from any male. The mask hid her smile, but he knew it was there. She was tall, about two inches shorter than his own six feet, with a figure that her modest dress couldn't hide.

"What can I do for you, Miss Lopez?"

"You remembered my name." Her voice was hushed.

"I try to remember my students' names." *Especially the attractive female students.* He looked at her with raised eyebrows, hoping she would get to the point. It was unfortunate, but he didn't have the time then.

Maybe he could give her a quick answer. "What can I do for you?"

She moved to his side, and they walked together. "Professor McGavock said you know something about the Stenhouse Field. I've been reading Doctor Stenhouse's paper and there are points I don't think I understand well enough."

"His research was groundbreaking. I can understand why you would be interested." He told himself he should brush Camila off; he had a test to study for. Maybe he could help her after that. It was hard to convince himself to put it off, though, knowing the smile behind the mask.

Camila nodded. "Stenhouse studied dark energy but found a field that could warp space without a gravitational source."

"Right. It didn't tell him anything about dark energy, and many scientists doubt that the field he discovered actually exists. His work suggested an answer to the puzzle of dark matter, though." Ben hesitated. "Did Professor McGavock tell you that my thesis is about the Stenhouse Field and its link to dark matter?"

"No. He just said you believed Doctor Stenhouse had really discovered something."

Ben smiled, but it was lost behind his mask. "Stenhouse's paper is almost everything known. To my knowledge, no one else has tried to replicate his results. That's part of what I hope to do."

Camila nodded. "He showed the field could warp space. Think about the applications for something like that. Spaceships that could travel faster than light, even."

"I don't think so. The field effects propagate at the speed of light. I don't see a way around that."

He could see the disappointment in those brown eyes and wanted to say something to restore the excitement that had been there. "Look, I've got a big test coming up. We can talk about this after that."

Camilla's eyes narrowed, probably because she suspected his motives. But she shook her head. "Sure. We can meet in the Student Union sometime next week."

Of course. Someplace public. Still, it would be pleasant to spend time with her.

B EN WALKED BACK TO his apartment, thinking about Camila. But her interest in him was probably only academic, and eventually he turned his attention to his Monday test on quantum field theory. Ben didn't think much about Camila's comments regarding the Stenhouse paper until Tuesday afternoon.

Camila had speculated about using the Stenhouse discovery for space travel. The technology wouldn't be useful in building starships, but what about travel within the Solar System? A vehicle that could reach even a reasonable percentage of light speed would make an enormous difference in support for the struggling Martian outpost.

But he saw the United States's effort on Mars as an expensive overreach. The world had other critical issues to deal with. Global warming, for example.

Stenhouse had proposed that dark energy was a field that didn't react with ordinary matter but did affect gravity. He claimed his experiments with what he called the Stenhouse Field showed space could be distorted without requiring a massive object, but had shed no light on the nature of dark energy. Camila suggested that, like so many other scientific advances, that ability could be harnessed. It seemed farfetched, but it was possible she had an idea worth looking into. At least he could spend time with Camila.

"I've given some thought to what you said about Stenhouse's paper," Ben told Camila as she entered the classroom. The class was a discussion period led by Ben, not one of Professor McGavock's lectures. "Are you free after class?"

"I have a class at three," she responded. "We could go over to the Student Union and talk about it before that."

The Student Union again, the second time she suggested it. Technically, there was no reason a student couldn't meet in public with a teaching assistant, but there would be talk if someone saw them together. So what?

He nodded. "Sure, the Student Union sounds good."

She smiled, perhaps a little smugly, and took a seat. Ben started the class, but he had difficulty concentrating on the material he was teaching. After class, they walked to the Student Union, got coffees from the service counter in the canteen, and sat down.

"You didn't think much of my idea last week," Camila said. "Have you changed your mind?"

"No, but we should talk about it more. I don't see how a Stenhouse Field could be used to break the light speed barrier, but that doesn't make it useless. Suppose propulsion systems could warp space enough to reach even a quarter of the speed of light. Supply ships could travel to Mars in hours instead of months. Asteroid mining could become practical." Ben had seen excitement in her eyes before, but without the mask, it transformed her entire face.

"That sounds great. Maybe you could help me understand some parts of the paper I had trouble with."

"I can try. Do you have Stenhouse's paper with you?"

Camila pulled a reader from her purse and opened the screen. "Sure." She entered a command and showed the result to Ben. "The field acts against gravity. Could it be used to reverse it? Anti-gravity."

Ben had read the paper many times, but he took a few minutes to reread it while Camila sipped her coffee.

When he finished, he handed the reader back to Camila. "The field warps space near the source, but then the effect drops off sharply, like the strong force, creating what amounts to a boundary. I suppose it might be possible."

Thirty years earlier, the world had finally begun addressing the dangers of global warming. The United States government, hamstrung by politics, was slow to react and stingy in its reaction. Instead of it leading the fight, other nations took over.

Advances were made in renewable energy, mainly in the developed world. Some people, wishful thinkers in Ben's opinion, said the problem was solved. But the infrastructure wasn't there for much of the world, and many countries still burned coal, wood, and petroleum. There were other continuing problems too: potable water, arable land, and health care, for example. All were worse or resistant to solutions in part because of climate change, and all needed available, inexpensive energy to fix.

Camila was staring at him. He had been silent too long. "I think it could also be used to increase gravitational force. If that's true, it could be a power source, used to trigger fusion. Maybe this could be the cheap, non-polluting energy we need."

Camila rested her elbow on the table and cradled her chin with her hand. "I suppose."

"This is your idea to run with. Write up a thesis proposal and see if you can get it accepted."

"My idea was to build spaceships."

"If people like Senator Correia have their way, global warming could make the expense of space exploration kill it before it gets anywhere. A source of plentiful, clean energy would go a long way to preventing that. It would make the developer a lot of money, too, that could be used to push your spaceship idea. Think of SpaceX and how that made near-Earth space travel practical."

"It would be years away,"

Ben nodded. "Better years away than not at all."

Camila was silent, and Ben could see she was considering what he said. He let her think. He didn't want to be seen as pressuring her, and after a minute, his approach was confirmed.

"Would you help me, especially at first?" Camila asked.

"Of course."

Ben met Camila again at the Student Union on Friday after classes. They ordered beers and sat down to talk, but their collaboration almost ended ten minutes into their conversation.

"Is it practical to cause fusion by manipulating a Stenhouse Field?" Camila asked.

"Possibly. Using warping to compress a gas until the temperature rises enough to cause fusion."

"Right."

Ben took a gulp of his beer and stared at the ceiling. He tried to consider the idea, and he could think better if he wasn't looking at Camila. "That would need temperatures of millions of degrees Kelvin you would have to control. Unless you can get a hell of a lot of compression, you would still need lasers to get the fuel hot enough."

Camila shrugged. "If it were easy, someone would have done it by now. At least, that's what my boyfriend says."

This was the first time Ben had heard anything about a boyfriend. He told himself it didn't make any difference; she had no interest in him other than what he could tell her about the Stenhouse Field. The jealousy that suddenly clouded his judgment was irrational, or so he told himself. A part of him wasn't listening.

He wanted to put down her casual remark, but held it in. He had that much control, at least. Perhaps he should stop wasting his time and let Camila and her boyfriend deal with her project on their own.

Camila stared at him. "Anything wrong?"

Ben forced himself to focus. *I'm being stupid.*

"No, I was just thinking about your suggestion about fusion. You should research that idea and see where it leads."

"OK. What do we do first?"

"You need to know a lot more about nuclear fusion and attempts made to use it if this is going to be your thesis."

Camila frowned. "That might be a problem. My parents are paying for my bachelor's degree, but I don't think they could handle grad school. It's too late to get a scholarship, and a summer job won't be enough."

"What about a loan?"

She wrinkled her nose. "At the interest rates they charge? If this idea doesn't pan out, I would be in debt for life."

Ben nodded. "I get a discount on my tuition and a small salary as a teaching assistant. My parents cover the rest. Could you do that?"

"Maybe. That might make a loan smaller and more practical."

Apparently, any more help from her parents is out of the question. He smiled, hoping to encourage her. It worked, but not the way he intended.

Camila smiled. "Maybe it could be your thesis project. If you're right about how it would help the environment, it would certainly be worthwhile."

"I'm already committed to my thesis project." He looked up again, wanting to avoid seeing her reaction.

"I don't know if I can."

He could hear the disappointment in her voice. *Does she think I would upend my life just because she has an idea?*

He kept his gaze above her head. "I can help, up to a point. But this is your project. Maybe your boyfriend will help."

Camila didn't answer, and after a few seconds, Ben dropped his eyes. She stared at him, and he knew he had made a mistake.

He tried to recover. "I'll help. You have time, though. For now, you're just getting together enough information for a thesis proposal."

"Assuming I go for a graduate degree and need to write a thesis."

The conversation was getting less awkward again, and Ben felt more confident. "If you're serious about exploiting Stenhouse's work, you must. If you're not serious, we might as well not waste time talking about it."

Camila frowned. "I'm serious about it."

"Then get going on the proposal. You'll need it to apply for a graduate program."

IN CAMILA'S EXPERIENCE, BEN was a typical male. She hadn't missed the way his face changed when she mentioned a boyfriend. Any relationship other than academic was inappropriate, and neither had expressed an interest in one, but Ben still reacted negatively to the idea there might be another man in her life. She shook her head. She didn't need that kind of drama.

Her roommate, Danielle Howell, was on her phone talking to her boyfriend when Camila got back to the apartment they shared. Dani had gotten her bachelor's degree the previous year and was now working for an MBA. Her boyfriend, Mike Terry, had already earned an MBA and worked at his father's investment company.

Dani waved to Camila and talked to Mike for another minute before ending the conversation. "So, how did things go with your new boyfriend?"

Camila frowned. "He's not a boyfriend. He's just helping me with an idea I had."

"Greg called a little while ago. He has tickets to the concert at the Field House tomorrow night, and you haven't told him you could go yet."

Camila had told Ben she had a boyfriend, but after two dates, she couldn't consider Greg a boyfriend, really. "I should study," she said. "I'm not a fan of Scarlet Sound, anyway."

Dani smiled. "And not a fan of Greg either, I take it. So, is this teaching assistant going to be a replacement?"

Camila shook her head. "There's no connection, at least romantically, and a relationship with an instructor is a complication I don't need."

"So why are you dating him?"

"It wasn't a date. He's smart and knowledgeable. He can help me with this idea I have, probably more than anyone else I could find."

"OK. I'm still a little fuzzy about this big idea of yours. Something about making spaceships, right?"

"That was my original idea. Ben thinks it would be better to develop a way to use the technology as a cheap, clean power source."

"He's probably right. We need that more than we need travel to other planets." She pointed to the mask now laid on Camila's desk. "Maybe we could get the air clean enough to get rid of those."

"I suppose. Ben thinks I should write a thesis proposal to exploit the discovery. That assumes I can go for a graduate degree, though."

"Why wouldn't you?"

"My parents sacrifice a lot just to pay for my BS. I don't think they can afford a graduate degree." She didn't bother mentioning that her brother would graduate from high school in another year and would also need money for further education.

Dani nodded. "Mike's father is paying for my MBA, since my hockey scholarship only covered my economics degree. Then I'll go to work for his father, too."

"I don't think Ben or Greg have rich families." Camila sighed and brought one of her textbooks up on her workstation. "Maybe I could wait a year and get a scholarship for the 2049 classes."

D ANI HAD LISTENED WHEN Camila tried to explain her idea, but terms like warped space were only vague memories of boring television shows that Camila sometimes watched. Camila talked about how some discovery could revolutionize space travel, and Dani paid little attention. Now, though, she talked about energy, and Dani understood that.

Camila was smart, smarter than she was. Camila had gotten into college with her brains and Dani, while hardly dumb, had gotten there on a hockey scholarship. Was her roommate on to something? It was probably a long shot, but if Camila's idea was a tenth as good as she seemed to think it was, it would be big.

Mike would do anything for her, but his father had all the money. She might improve her standing with Mike's family if she could convince Mike's father that a small investment would be worth the risk.

G REG MORA SLAMMED THE phone down on his desk and swore. So Camila had to study and couldn't see him that weekend. He'd already bought the tickets, and now he was stuck with them. Admittedly, he probably should have asked her first, but the concert might have been sold out by the time he had the chance. It would have been their third date, and he had hoped it might result in more than a quick kiss at the entrance to her apartment building.

He had noticed her in the Student Union with some grad student. They were only talking, and he wasn't some jealous child who couldn't stand to see his girlfriend talking to another man, but now he wondered if he was too understanding. Maybe he should have followed his first instinct and joined them.

She certainly had given him the impression she was as interested in him as he was in her. Her smile would have gotten his attention even if she wasn't the best-looking girl he knew. But he wasn't a graduate student, and perhaps that made a difference. He swore again.

He could try to compete with this other guy. The bastard looked like a teek, and Greg was more athletic, but Camila was a physics major, and maybe teeks were her type. He could find a girl who appreciated what he had to offer, but, damn, Camila was so hot.

DANI AND MIKE TERRY went to the concert that Camila passed on. After the show, they went to their usual place, a bar frequented by university students. The place was crowded and noisy, but Mike tipped the host. The host, another college student, accepted the money, smiled, and escorted them to a booth in one corner, repeating a ritual that occurred on most weekends.

Dani slid into the booth and Mike followed her, moving in just enough for a sexy closeness. He kissed her; she kissed back and let him put a hand on one of her breasts. Another often-repeated weekend occurrence that she hadn't let get any further.

"You remember my roommate, Camila?" she asked when Mike paused to catch his breath.

Mike halted his bid for another kiss and looked at her with a frown. "Yeah, I think so. Why?"

Dani suppressed a laugh. As if Mike didn't remember Camila very well. "She's a physics major. She was talking to me about an idea she had."

"OK. Why are we talking about Camila or her idea?"

"It was interesting. She was talking about a clean way of generating power, and it might be worth some money."

Mike's frown deepened. "Sure, if there's anything to it. I'll bet every physics teek has an idea like that."

"Maybe. But she's smart and has one of the graduate instructors interested, too. It might be something you could tell your father about and maybe impress him."

"Who's the grad instructor?"

"I think his name is Ben Pinto. Do you know him?"

"No, but I have friends that probably do." Mike leaned toward Dani for another kiss. "Let me ask around and see if there's anything to it."

Dani smiled, put a hand behind his neck, and pulled him into the kiss, putting a little extra into it. "Thanks, Mike. It might be something important." Meaning profitable.

Ryan Terry looked up as his son, Mike, came into the room. Ryan's full name was Terence Ryan Terry, but his first name was a family secret, seldom mentioned against the possibility of becoming known as Terry Terry.

"Good morning," Ryan said. "Come home to get someplace quiet to study?"

Mike grinned at him. "Maybe later." He sat in the chair closest to the desk where Ryan was examining financial reports on his workstation. "Mostly, I came over to talk to you about an investment opportunity."

Ryan looked at his son. "Is that what you young people call it now when you want to get money from your parents?"

"No, nothing like that. It wouldn't even be for me. I know this physics major with an idea that could be really big."

"No matter how many fancy words they use, a perpetual motion machine is still impossible."

Mike chuckled, although Ryan suspected he faked it. "I know that, but this might be almost as good."

"You said this is from a physics major. So not your girlfriend?"

"Dani's roommate, actually. Camila has a grad student working with her, too."

Ryan sighed. "OK, what's this big idea?"

"Some guy named Stenhouse discovered a way to warp space. They think it could be used to create green energy."

Ryan remembered hearing something about Stenhouse's discovery, but his expertise was in finance, not science. Warping space was something from science fiction, used, at least fictionally, for space travel, not power generation.

He turned back to his workstation. "Give me information on Stenhouse and warped space."

"Arthur Stenhouse is the lead scientist at the Advanced Physic Institute in Spokane, Washington," the computer responded. "His paper, 'The Manipulation of Dark Energy,' did not advance knowledge of what dark energy is, but a subsequent paper suggested the field reported in the earlier work was the source of dark matter. He speculated what he called the Stenhouse Field might be useful for space exploration."

"How would that be used to get energy?" Ryan asked Mike. "This guy Stenhouse is talking about spaceships."

"I don't know, but Dani's roommate said warping space could be used to generate power, too."

"So, how much money are we talking about for this investment?"

"Camila can't afford to pay for the graduate degree she would need. Dani suggested you might do it to get in on the opportunity."

"And helping her roommate would help you with your girlfriend."

Mike grinned. "Win-win."

"All right. No promises, but I'll think about it."

AFTER MIKE LEFT, RYAN tried to get back to his financial reports, but found that difficult. Mike's proposal was probably nonsense, but the availability of energy drove the economy, and he couldn't ignore the possibility that there was something behind the idea. An innovative technology, developed in the United States, could do a lot to reverse the decline of the country's influence in the world and make its developers rich.

Not enough had been done about global warming. In the previous two decades, actions by governments slowed the accumulation of damage, and the United States and other developed countries declared victory. They reduced spending on mitigating the problem, but for many scientists, that was premature.

If Mike's friends had a realistic idea, industry might solve what governments no longer cared to do. A solution would be immensely profitable. Of course, that all depended on the idea being practical.

His son wasn't stupid, but he was a typical male of his age, more interested in women and sports than science or making money. Mike's girlfriend, Danielle Howell, although a player on the school's women's hockey team when Mike met her, was different. Ryan recognized something of himself in her: aggressive, focused, and manipulative. Ryan believed she genuinely cared for his son, but his money was an important aspect of her feelings for Mike. If she saw merit in the proposal, he should probably take it seriously.

But he didn't have the expertise to evaluate the opportunity. He needed to know more about the two students who had come up with the idea, too. Later, over lunch, he asked Mike if he knew for whom Ben Pinto worked as an assistant.

"Professor McGavock," Mike told him. "He teaches astrophysics."

Ryan nodded. Talking to Professor McGavock might help him decide how to proceed.

R YAN RARELY FLOUTED HIS success, but his instincts told him that a show of wealth would impress the academic. For his appointment, he wore a recently tailored, expensive suit that complemented the body he cultivated in his well-equipped home gym. A Rolex watch, tasteful jewelry, and perfectly shined shoes from a fashionable, pricey Italian designer completed the look.

Doctor Anthony McGavock, although about the same age, was different. The teacher wore a rumpled sport coat over a dress shirt open at the collar. His abundant blond hair hadn't been cut in a while and a scar on his left cheek implied an active past. The eyes below bushy eyebrows were impatient.

"I would like to ask you about one of your students," Ryan said after they sat down.

"Of course. Mr. Terry, right? I don't recall a student with that name."

"I'm not her parent. But my son is dating the roommate of Camila Lopez, and he told me about her. I'm interested in your teaching assistant, too. Ben Pinto."

"I don't understand. Camila is very bright. So is Ben." McGavock gave Ryan a hard stare. "Camila is attractive, too. What's the nature of your interest?"

"According to my son's roommate, Camila and Ben are developing an idea for generating clean power. Something about a better way to cause fusion. If there's anything to it, it could be enormously profitable, and I'm considering helping them out financially."

"I know nothing about that."

"Would they know the difference between a pipe dream and an idea with real promise?"

The professor scratched his head. "Ben probably would. Anything helping the environment would attract him, but I don't think he would chase some fantasy. I don't know Miss Lopez very well. She's an excellent student and she will graduate this year, but that's all I know."

"Their idea involves papers by a scientist named Stenhouse. Something about dark energy and warped space."

McGavock chuckled. "She asked about Stenhouse, and I sent her to Ben. He's doing his thesis on an aspect of Stenhouse's work, but I suspect he'll just prove that Stenhouse was wrong. Stenhouse was trying to explain dark energy, you know, but failed. Dark energy is uniform throughout the universe, unsuitable for extracting energy." He peered at Ryan curiously. "I don't know how this would relate to fusion. Can you tell me more about their idea?"

Ryan didn't like the look on McGavock's face. *I shouldn't tell him too much.* Ryan had heard stories about scientists stealing the work of their subordinates, and he wanted to be the one to take advantage of the two students. As he helped them, of course. Anyway, he probably didn't understand enough about it to tell the professor anything.

"That's why I wanted to talk to you." He waved a hand. "I'm a financier, not a scientist. I don't understand most of what I was told. If you think it's a dead end, you're probably right." He paused. "Thank you for your time, Professor."

PROFESSOR McGAVOCK STARED AFTER Ryan. The astrophysicist was one of many scientists skeptical about Stenhouse's claims and had tried to dissuade his assistant from basing his doctor's thesis on Stenhouse's work. But Ryan Terry's expensive attire marked him as a man who didn't support foolish ventures. Could they be right?

It wouldn't hurt to attach himself to the project. They were both his students, and Camila Lopez would need a thesis advisor if she moved ahead. Now he regretted his discouraging words to Terry. But perhaps Lopez's charm could still win him over.

RYAN TALKED TO HIS son, and Mike had Dani set up a meeting with Camila and Ben at a downtown pizza restaurant. Ryan remembered his own college days studying finance and economics and assumed science students would appreciate a free meal as much as he had.

"Your roommate has been telling my son about your idea," he told Camila while they waited for their food.

Camila looked at him and smiled. *She may be a science major, but she knows how to use her appearance to charm.*

"Are you planning to do anything with it?" Ryan included Ben in his gaze.

"Ben and I have discussed it," Camila said. "We're just students, though. Ben thought I could use the idea as a thesis topic if I went on to an advanced degree."

The problem with that approach would be the years it would take before she was ready. His talk with Professor McGavock left the impression that the idea wasn't practical, but his son had brought it to him. That counted for something.

"That sounds like solid long-range thinking," Ryan said. "As Dani has probably told you, I have been quite successful at finding profitable investments. She told me you're not sure you can pay for a master's or doctoral degree."

Camila nodded. "Just paying for my BS is a stretch for my parents."

"What about a scholarship or a loan?"

"I didn't think an advanced degree would be possible, so I didn't apply for any scholarships. I could skip a year and perhaps get one next year." Camila paused. "At the interest rates being charged, I don't think it would be smart to borrow too much. Ben suggested I become a teaching assistant and take out a smaller loan."

"How much interest are you talking about?" Ryan asked.

"If I'm lucky, ten percent. Probably higher."

Ryan had lost money on investments before. Most serious investors balanced the losses with gains. But the conversation showed him a way to get involved with the two students with minimal risk, but still look good to Mike.

"I could lend you the money to pay for an advanced degree. Would that work for you if the interest rate was more like, say, four percent?"

"What's in it for you?" Ben asked. He looked at Ryan suspiciously, while Camila's face showed only surprise and perhaps excitement. Ben was clearly the more practical of the two.

"If you really have a promising idea, we would all benefit. The only downside for me is that I would be trusting you to pay me back."

"Couldn't you make more by investing in something else?" Ben asked.

"Maybe. Or I could lose money. This would be a small, but relatively safe investment."

He could almost see the wheels turning in Ben's brain. It was a no-risk situation for him. He could spend time with Camila, a plus for any normal male, and he stood a chance at becoming rich. From what Mike had told him, the possibility of helping the environment would also help convince him. After a few more seconds, Ben proved him correct.

Ben looked at Camila and shrugged. "It sounds good to me, but you'd be borrowing the money."

Camila looked at Ryan, and he could see she was interested. "Can I think about it?"

"Of course."

"If you're sticking around for grad school, I don't have to find another roommate," Dani told Camila. "Mike is hinting I should move in with him when you leave." She snorted. "Like that was going to happen. My parents are old-fashioned about things like that."

"I'm still thinking about Mr. Terry's offer, but if I accept, I don't see why not." Camila chuckled. "I didn't think you worried that much about what your parents thought."

Dani shrugged. "OK, you've got me. I need to keep some leverage on Mike. With the restrictions on birth control and obstacles to abortion, this is no time to get pregnant. I don't want him to think he can skip the marriage part."

Camila looked at her, and Dani saw a slyness in her expression. "I don't suppose the Terry money has anything to do with that," Camila said.

"It's a nice bonus. I do love Mike, though."

Camila drew back and nibbled on her lip. "I appreciate you introducing me to Mike's father. His financial support would be such a blessing."

"When will you decide?"

"I want to talk to my parents first. But unless they can come up with a good reason not to, I think I have to accept his offer."

Camila was in New York, but Texas was home. That night, she called her parents, Tomás and Nina, and told them about the loan offer.

"We could manage it," Nina said. "You don't have to borrow money."

Tomás appeared on the screen and nodded, but Camila could see doubt in his expression. "It's a good deal, Poppa. Not like those high interest rate loans. You know I appreciate the sacrifices you've made to get me this far, but you've done enough."

"That's not possible for my little girl," Tomás said.

Camila smiled. "I'm not so little anymore. And Diego will start college soon. I think this is a wonderful opportunity, but I wanted to get your opinion before I agreed."

Tomás frowned. "Just make sure you have a written agreement and that you read it carefully."

"I will, Poppa. You've done so much for me. I hope I can pay you back some day."

"Study hard, darling," Nina said. "We're very proud of you."

B EN HELPED CAMILA CALCULATE what she would need, and Ryan supplied a standard loan agreement. Payments wouldn't begin until after she graduated.

That was the easy part. Camila had to write a thesis proposal while continuing her undergraduate work. Although she could now afford an advanced degree, she still had to be admitted into a program that had limited enrollments available. Ben helped guide her through the process.

He helped her with her thesis as well. His doctoral thesis also researched the Stenhouse Field, but was more theoretical. Both would have to design and build a device to generate the field, but Ben's needs were closer to Stenhouse's original research, unique in his plan to study the field in zero gravity. Camila would need a stronger, more controlled phenomenon.

He was still a teaching assistant for Professor McGavock and met with McGavock each Monday morning, a brief discussion of any issues that might affect the coming week. At their next meeting, McGavock brought up his collaboration with Camila.

"I've heard rumors you're seeing one of my students." McGavock peered at Ben from the other side of the desk. "Camila Lopez?"

"Yes sir. I'm helping her with her application to graduate school."

"She's a very attractive girl. I hope nothing inappropriate is going on."

"No sir. We're working on an idea she had. Something she hopes might make a good thesis project when she goes for a graduate degree."

"Yes, Ryan Terry said something about that. Another investigation of Stenhouse's work with dark energy, I believe."

Ben was surprised, but it would be natural for Terry to investigate Camila before committing his money. "Yes. I think she has a promising idea. Mr. Terry agrees and will pay for her graduate degree."

McGavock's eyebrows rose. "That's rather generous of him, considering most scientists don't accept Stenhouse's work."

Ben wavered in responding, caution taking over. "He's only lending her the money."

"You should consult with someone with more experience." The professor smiled. "I'm sure I could find time to help. As an investment of my own."

"That's very kind. I will mention your offer to Camila."

McGavock nodded. "Good. I look forward to working with you."

R YAN MET WITH CAMILA and Ben at a pizzeria near the campus to exchange status information. It wasn't necessary, but he enjoyed listening to the discussions, even if he only understood half of what the two students said. Camila reported on her thesis proposal, and she and Ben tried to explain with limited success their progress toward understanding the Stenhouse Field and nuclear fusion.

Ben told them about his conversation with Professor McGavock. To Ryan, it sounded like the academic was maneuvering to take advantage of his students. Perhaps he wasn't as doubtful about the idea as he had implied in talking to Ryan. But even if Camila's ideas worked out, exploiting them would be a major engineering project.

His ex-wife had maintained money wouldn't buy love, but he didn't know any other way to show his love for Mike. So, when Mike asked for something, even support for a friend, Ryan usually came through.

"Can this professor help you?" he asked.

"His expertise is in astrophysics," Ben said. "If we need someone, there are a lot of scientists who could help more. Camila will need a thesis advisor, probably one in nuclear physics, and that person will be more helpful. My faculty advisor, Professor Moreno, would be a better choice."

Still, McGavock could cause trouble for his students. "Thank him for the offer," Ryan told Ben. "Tell him it's too early to be thinking about bringing him in, but that we would keep him in mind for the future."

HER CLASSES WERE OVER for the week, and Camila had a little time before dinner. She opened the notes on nuclear fusion on her workstation, hoping to spend the time thinking about the problem of compressing hydrogen enough to initiate fusion. The sun used the gravity of its immense mass to force hydrogen atoms to fuse into helium, and the half-dozen fusion power generating plants in the world used pressure and lasers to create the necessary plasma state. She wanted to do it with compression alone, and she needed to demonstrate how she proposed to do that to get her graduate application approved.

She brought Stenhouse's second paper up on her computer and scanned it, hoping for inspiration. A short paragraph about an observed faint glow caught her attention. She hadn't paid much attention to it before; Stenhouse speculated it was a minor phenomenon at the edge of the field and then ignored it.

Her phone announced a call from Greg, and she frowned. She had gone out with him twice more, but his increasing pressure to sleep with him was annoying and a bit repellent. Although Dani said no, she might be having sex with Mike, but they were committed to each other, and Camila had no interest in a long-term relationship with Greg. Dani was better company.

"Hey Cam, let's get together tonight," Greg said when she answered.

Camila glanced at the computer screen. "I don't think so." The words on the screen interested her more than Greg did, and she needed to make more progress on her thesis proposal.

"What, you have something better to do?"

"Yes."

Greg was silent for several seconds. "You're not trying very hard to make this relationship work."

"You're right. I guess I'm just wasting your time, Greg. Why don't we end this so you can move on?"

"Just like that? Come on, Camila!"

"Sorry, Greg. We tried, but I can't give you what you want and, to be honest, you can't give me what I need."

"But Ben Pinto can?" Camila could hear the anger in Greg's voice.

"I need Ben for his knowledge, so yes, I guess he can."

She heard a bang. Greg had probably pounded on his desk or some other surface. Then the connection was broken.

She shrugged and got up to make herself some lunch.

E VEN FOR SOMEONE WITH Camila's grades and the assurance of financing, acceptance into a postgraduate program wasn't automatic. She polished her thesis proposal, but still needed a thesis advisor. Ben recommended his own advisor, Professor Victoria Moreno. Camila had taken her introductory course in nuclear physics the previous year, but hadn't seen her since.

On Tuesday morning, she met with the teacher in her office. Professor Moreno greeted her warmly, and they sat together on a couch in one corner of the office.

"I was pleased to receive your proposal," Moreno said. She was probably in her sixties and would have been undistinguished in appearance without the stylish eyeglasses perched on her nose. With the corrective medical options available, few people wore glasses, but Moreno had once confessed to her class that she wanted to look more professorial.

"Thank you, professor."

"You were one of my best students. I'm so happy you're continuing your education. Obviously, I would be delighted to mentor you. Are you going for a master's or the combined program?"

The school had created the combined program five years before, condensing the class requirement for a doctorate slightly to three years, with research usually performed afterwards. It was not a time-saver as much as a declaration of intent that would avoid a second application after getting a master's degree.

"The combined program. I know I'll need a PhD."

"Your thesis topic sounds interesting. 'The application of Stenhouse's work to the exploitation of nuclear power.'"

Camila nodded. "Yes, I hope to show how it could harness fusion much easier than in current facilities."

Moreno sat up in her seat and her face split into a wide grin. "Wonderful! He's had trouble getting his work accepted, but I think his discoveries will make him one of the greats. He should have already gotten a Nobel." Her face became more serious. "But he didn't think his work could be exploited to produce energy."

"Not directly, but I'm working on the possibility of using warped space to push atoms together with enough force to start fusion. My research would determine if that is practical."

"Yes, your proposal was interesting. I'm not sure the review committee would agree, however." She rose and went to the computer on her desk. "I have your proposal here. It's good, but still weak on how you plan to generate enough pressure to initiate fusion. That will be a problem during review."

Camila bit her lip. "I know. If I need lasers to raise the temperature, it won't be an improvement over existing methods."

"Exactly." Moreno came back to the couch. She put a hand on Camila's knee and Camila found her smile encouraging. She remembered thinking about the glow Stenhouse had seen. Greg's call had interrupted her thoughts on that.

"There is something else I could look into," Camila answered. She explained the effect, and the teacher went back to her computer and displayed Stenhouse's paper.

"Something is causing a release of energy," Moreno said. "You're working with Ben on this?"

Camila nodded, and Moreno continued. "You should look into this more. Ben's work might offer some insight. Don't forget about fusion yet, but broaden your scope."

"You think that would improve my chances of being accepted?"

Moreno nodded. "And your chances of discovering something important."

T HAT WOULD MEAN SOME work revising her proposal, but Camila couldn't ignore Professor Moreno. After leaving the teacher, she called Ben and arranged a meeting for that afternoon.

"I saw that," he told her when she mentioned the glow. "I didn't think I could use it to prove the field existed. It's quite faint."

"But doesn't it mean energy is being produced? It may not be relevant to proving the field exists or studying dark matter, but it could be important for my thesis."

Ben rubbed his chin. "Stenhouse generated the field in a vacuum, but no vacuum is perfect. There would always be a few atoms. Stenhouse might have assumed that they were being ionized."

Camila didn't agree. "I don't think that would explain it. We know the field affects gravity. Maybe it affects the strong force as well."

Ben raised his eyebrows. "You think it's breaking up nuclei? That would release a lot of energy."

"But only for elements heavier than iron, right? There wouldn't be many of them."

Ben pulled his reader and entered a series of commands. "I guess it's possible. Worth researching, certainly."

S HE SPENT THE NEXT two days revising her proposal. With the submission deadline three days away, she met again with Professor Moreno.

"Concentrating on fission possibilities should help," Moreno said. She rubbed her chin. "Fusion research could have been really expensive, but I think investigating the glow could be done with a reasonable budget."

"So you think they'll accept my application?"

"I hope so." She entered a command on her computer. "By the time you begin your research, Ben will have produced more evidence that the field is real, but you don't want to wait for that. The three professors on the committee will be Daniel, McGavock, and Curtis. Dean Curtis is the most conservative, I think. He got his degrees at the peak of string theory popularity."

"Professor McGavock?" Camila repeated.

"Yes. He's the youngest of the three and perhaps the most open to new ideas. Why?"

"I'm taking a class from him now, and Ben is his assistant. In fact, Ben recommended you as my advisor."

Moreno nodded. "Is Professor McGavock a problem?"

"I'm not sure. He has offered to help, but he's skeptical of Stenhouse's work. I've politely turned him down so far."

"Maybe he will at least help you with the review committee."

"Maybe."

O N FRIDAY NIGHT, THE group got together at the pizzeria where they had met before. "I'll be submitting my thesis proposal sometime next week." Camila drummed her fingers on the table, realized she was doing it, and stopped with a frown.

"They'll accept it," Ben said. "It's a solid proposal."

"That's not what I'm worried about," Camila said. "Professor Moreno thinks the topic might be too controversial and the research too expensive."

"Why would nuclear energy be controversial?" Ryan asked.

"Not nuclear energy, really, but not all scientists accept Stenhouse's results." Camila shook her head. "I never paid much attention before, but apparently academic politics can be as senseless as government politics."

"And Professor Moreno is part of that?" Ryan asked.

"No, not her. She's a fan of Stenhouse. But she's afraid of what the board might think."

"It might not hurt if I dropped a hint with Professor McGavock that you might need some of his help if the board accepts your proposal," Ben said.

Ryan leaned forward. "What does he have to do with it?"

"He's one of the three professors on the review committee," Camila answered.

Ryan nodded. "Hint, but don't promise anything."

Mike changed the subject to the women's hockey game, one of Dani's favorite dates, for the following night, and they didn't bring up the thesis again. Whenever Camila looked at Ryan, though, he appeared deep in thought.

T HE INTERCOM BUZZED FOR his attention, and Kaleb Watts, the president of the university, looked up from the financial report he had been reading. "A Mr. Ryan Terry is here to talk to you about a possible endowment," his secretary said.

"Show him in." People with money rarely came to him, at least at first, but the school could always use an endowment. The name wasn't familiar to him, but that didn't mean the visitor didn't have a healthy net worth.

Watts studied Terry as he entered the office. He was probably in his late forties or early fifties, dressed to look like money. Of course, that could be an attempt to mislead, but that didn't happen often at Watts's level.

"How can I help you, Mr. Terry?" Watts asked as Ryan took a seat.

Ryan leaned across the desk toward Watts. "My son is friends with several students from this school and has talked me into loaning one of them funds for a doctoral program. She still has to have her thesis topic approved, but I'm sure that's just a formality."

"The student is a senior here now?"

"Yes. Camila Lopez. She's in the Physics Department."

"We welcome your generosity to one of our students, of course," Watts said. "I understood this was about an endowment, however."

"Assuming she is accepted into the program, the loan is a done deal. She loves this school, though, and is trying to convince me to make a significant endowment to it. I understand that there might be a problem funding her research, and I could help with that financially."

Watts smiled. "That would certainly be appreciated."

"I've been lucky financially and am always looking for ways to give back. Camila is very persuasive. I have wealthy friends who might also be persuadable."

They talked for a few minutes longer, but Watts was a busy man, and he was sure Ryan Terry was as well. They shook hands, and Ryan left.

Watts stared at the door and shook his head as Ryan Terry disappeared. *He's playing me.* Still, that didn't mean he shouldn't investigate a bit.

It took only a few keystrokes to verify that Camila Lopez was a senior in the Physics Department and that she had applied for the doctoral program. He saw the title of her thesis proposal, *The Application of the Stenhouse Field to the Exploitation of Nuclear Energy,* and nodded. Any research into another energy source would certainly be welcome.

Her thesis advisor was Doctor Victoria Moreno. Victoria got high marks as a teacher, and he was under the impression she was a competent scientist. He had overheard one or two comments about her being unorthodox, but that wasn't a terrible thing.

Next, he used the computer to research Ryan Terry. There wasn't much on him, but enough to suggest he probably could make at least a moderate donation to the college. That left only one question: what was Terry trying to do?

It could be an attempt to have Watts influence the thesis committee. Lopez was an excellent candidate, but physics wasn't his field. Another quick command told him who was on the committee and that, of the three, only Professor Daniel was immediately available.

"I've been asked about the PhD application for Camila Lopez," he told Professor Sarah Daniel when she answered the phone.

"We're reviewing her proposal next week," Daniel answered. "Is there a problem?"

"Not that I know of, but she apparently has a sponsor who is concerned. Is there a reason we wouldn't accept her?"

"I've read her proposal." She paused for a moment. "She did an excellent job on it, but the subject is unusual. Doctor Curtis won't like it much. The need for an appropriate research facility could be a stumbling block, too."

"Why would nuclear energy be unusual?"

"Her proposal to use warped space is unconventional. She's using research papers many physicists haven't embraced. I can go into it more if you wish."

"No, I'm sure it's over my head. There may be an endowment at stake, though."

"The sponsor promised to make an endowment if we accept her?" Watts could hear the indignation in Daniel's voice.

"He was more subtle than that. Miss Lopez allegedly urged him to donate, the implication being she might be less enthusiastic about it if she's not accepted. He suggested he might contribute to the cost of her research."

Daniel sighed. "I'll make sure she gets a fair hearing and doesn't get bullied by Dean. You might want to talk to McGavock, too."

"All right. Thanks."

P ROFESSOR MCGAVOCK KNEW THE university president, but not well. He was surprised but delighted when President Watts walked into his office.

"What can you tell me about Camila Lopez?" Watts asked after taking a seat.

"She's a student of mine. The committee is looking at her thesis proposal soon."

"Her benefactor visited me yesterday. He's paying for her PhD and is hinting that he might help finance her research. A facility might be a problem."

McGavock nodded. "Perhaps. It's difficult to know what she'll need. The technology she wants to work with is so new. But she plans to demonstrate only that a Stenhouse Field can generate energy. Possibly she could use an existing laboratory if she limits the output."

"It would benefit the university, though, if she made a breakthrough in the field."

"That's a big if, given the controversy about Stenhouse's results. But yes. Potentially a great deal."

"We don't have to lock in her thesis project immediately. She has three years of classes before she'll begin the actual research."

"In that time, if Terry delivers a large endowment, we could build a new facility." McGavock smiled. "That would be even better for us."

" C AMILA LOPEZ SHOWS A lot of promise," Professor McGavock said. "I don't think we need to spend a lot of time on this."

"Warping space sounds like a fantasy," Professor Dean Curtis responded. "Even if the idea has merit, this sounds more like an engineering subject, suitable when we know what we're dealing with. We need more evidence that this field even exists. And fusion research could be extremely expensive, requiring a new facility. I know Watts thinks she has friends who could endow one, but that seems unlikely, given the probable cost and time frame."

McGavock frowned at his colleague. Dean was in his eighties and probably should have retired long ago. The cottony cloud of white hair contrasting with his dark skin, the large glasses, and the grim expression did little to make him look younger. He had a reputation as a successful mechanical engineer before becoming an academic, but that was far in the past.

"We don't have a complete answer, but Stenhouse's work advanced our knowledge," McGavock said. "Camila's project could advance it further by researching the energy generation potential of the Stenhouse Field. It could be an enormous boost for the school."

"Camila?" Curtis snorted.

"Miss Lopez, then. She's a student of mine and has consulted with me on her ideas."

"This probably shouldn't matter, but President Watts talked to me, too," Professor Daniel said. "He left the impression it would be good for the school if we accepted Miss Lopez. And I agree with Anthony. She shows promise."

"It's not real physics," Curtis said. "It's engineering."

"If her research is successful, it could be a breakthrough in science," McGavock said. "Another student is basing his thesis in part on duplicating Stenhouse. If he succeeds, Lopez's project will be worth pursuing. If he fails, Lopez will have to submit a new topic."

Curtis shook his head. "All right, all right. The science might be suspect, and I have serious reservations about a facility, but her proposal is well thought out, and she seems to be intelligent. We'll approve her application, but I have reservations about whether her thesis work will be successful."

CAMILA GRADUATED IN MID-JUNE and went home to Texas. Ben, whose family lived in nearby Saratoga, stayed in his apartment. Camila hadn't talked much with her family during the last semester, and she looked forward to telling them more about how she would work toward a graduate degree. She discovered the details required some explanation.

"I don't understand why this man will pay for your advanced degree," her father said.

"He's not paying for it. He's only loaning me the money at a better interest rate than I could have gotten elsewhere. Fortunately for me, Mr. Terry's son is my roommate's boyfriend."

Tomás frowned. "Are you sure that's Mr. Terry's real reason?"

"Yes, Poppa. Don't worry. I'm never even alone with him. Ben is always with me when I talk to Mr. Terry."

"And what does Ben get from this?" her mother asked.

"He's helping me. That's all."

Tomás scowled. "What if Ben is after something else?"

"I can handle Ben." Camila chuckled. "He enjoys my company, I think, but that's all he'll enjoy."

"I thought you were seeing this boy Greg," Nina said.

"We broke up. He's the one you should have worried about. He was constantly pressuring me."

Camila's younger brother, Diego, had just finished his junior year of high school. "So, what's this big idea that might make you a billionaire?"

Camila was glad to get the conversation away from the irrelevant discussion of the men in her life. She launched into a long monologue about the project, stopping only when her mother interrupted to say it was time for dinner.

AFTER DINNER, DIEGO WENT to his room. His sister's talk about her project and how it might make her rich was interesting, but he wanted to start a new session of *A World of Your Own*. V-World, a company founded by his hero, Jürgen Varela, had released the game the previous year, and it had instantly become Diego's favorite.

Jürgen Varela had created a company around his concept of a virtual reality game that used artificial intelligence to generate a scenario on the fly based on input from the player. According to the company's press releases, he was a millionaire and still in his twenties. Why couldn't Camila do something like that? Diego was sure his sister was smart enough.

He wrapped game sensors around his wrists and ankles and started the program. The game's unique feature was its ability to generate worlds and scenarios as it ran, with unpredictable results. Virtual Reality stadiums used augmented reality to create a more immersive experience and had better quality graphics, but such facilities cost money and used fixed scenarios with limited variations.

"Begin new scenario using previously created world Hazard Planet. Character is a warrior male hunting dangerous aliens and armed with a blaster. Set difficulty to medium, action level high."

"Creating character Dirk the Mighty," the game said. "Scenario begins outside the village Serenity in the year 2450."

"Begin play," Diego said. After a brief delay, he stood at the edge of a forest of strange-looking trees. The walls of Serenity loomed behind him. He had created the setting before, but the game would generate a scenario based on his prior inputs and his starting prompt, developing the story as the game progressed.

Diego grinned. No matter how many times he played the game, he would encounter something new and exciting. Already, a large, toothy creature moved toward him from the trees, growling a challenge.

AT HOME, THERE WERE interruptions, but Camila found time to research the issues she would eventually have to deal with for her thesis. She talked with Ben every week and they compared notes, but they struggled with the problem of overcoming electromagnetic repulsion enough to force hydrogen to fuse into helium. Enormous pressures and temperatures were required for existing fusion energy production, and it wasn't easy to see how warping space could do it more efficiently. Investigating the glowing phenomenon at the field's boundary appeared more promising.

Summer passed slowly. Few people spent much time outdoors during the oppressive Texas summer, and maintaining more than minimal comfort indoors was expensive. When she became drowsy studying, she did a few household chores for her mother, a science teacher at the local high school.

The new term would begin on August twenty-fourth, and she flew back to school two days before. Her flight from Houston was scheduled to land a little after two in the afternoon, but someone had hacked into the FAA computers, delaying the takeoff for two hours. Even after the plane departed, residual problems with aircraft ground control increased the flight time. The problem seemed to affect ground traffic, too, somehow; the normal twenty-minute trip to the off-campus apartment she shared with Dani took her automated taxi over forty-five minutes.

When she entered the apartment, she was exhausted. Dani was out, so Camila put sheets on her bed, connected her computer, and shelved the few physical books she had. By the time Dani came back, Camila was lying on a couch, almost asleep.

"Ben called earlier, asking when you would arrive," Dani told her. "He wanted to get together to discuss some ideas he had."

"He could have called me in Texas." Camila yawned. Spending the rest of the evening with Dani sounded better than a conversation with Ben.

W HEN MONDAY MORNING CAME, Camila had other concerns. Classes wouldn't begin until the next day, but first, there was registration to deal with. That all should have been routine except for selecting classes and fitting them into a workable schedule. Camila had never liked that process, but students could usually complete it in less than an hour.

Many of the school's teachers were at the registration to answer questions. She saw Professor McGavock and tried to avoid him, but he found her.

He approached her as she was reading the description of a course she was considering. "It's nice to see you again. I'm glad I could help you stay with us and get your PhD."

"Yes, thank you." Camila glanced back at the screen displaying the course description.

"Well, I'm sure you're busy," McGavock said. "But the offer still stands. If there's any way I can help you further."

Camila smiled. "Thank you. I'll definitely keep that in mind."

10

ONDAY NIGHT, CAMILA RELAXED with Dani in their apartment. She had navigated registration successfully, and classes wouldn't begin until the next day. Dani's boyfriend, Mike Terry, was on a business trip for his father. Camila briefly considered inviting Ben to join them, but Dani agreed that a girls-only night sounded better.

Camila mentioned Professor McGavock approaching her during registration. "It's weird," Camila said. "I don't understand why he keeps after me when he has no interest in Stenhouse."

"I'm guessing McGavock was trying to get a piece of whatever you and Ben do, just in case you succeed," Dani said. "He's offered before?"

"When Ben and I first talked about it, he offered his help. Mike's father thought the same as you."

"There you are! If your thesis project works out, he could get a lot of prestige just by being involved."

Camila nodded. "I suppose. Professor Moreno will be more helpful, and she thinks the Stenhouse Field is real. I wish he would leave me alone."

"Well, if he gives you any more trouble, just threaten to report he made sexual advances, and that's why he wanted to work with you."

"I couldn't do that. Who would believe it?"

Dani grinned. "Anyone who has seen you, sweetie."

T THE PIZZERIA THAT was becoming their usual meeting place, Camila and Ben sat on one side, Mike and Dani on the other, with Ryan pulling up a chair on the aisle end of the table. Camila spent her time on her classes rather than her thesis, but Ryan had suggested they all get together every couple of weeks to talk, anyway. She, along with the others, enjoyed the companionship and appreciated the free pizza.

"So you're graduating this year, Ben," Ryan said while they waited for their food.

"My classwork only," Ben answered. "I'll start my thesis research this summer."

"I see. What's your project?"

"Stenhouse Effects on a Gravitational Field Explaining Dark Matter."

Ryan blinked several times. "OK, that sounds impressive to a mere economics major like me. I thought I knew a little, though. You were planning to perform experiments in orbit. But there's no gravity there."

"Not exactly," Ben said. "Gravity is always there, but in orbit, you're in free fall so you don't feel it. Many physicists feel the Stenhouse field is only an unexplained fluctuation in the gravitational field, and by duplicating Stenhouse's experiments, I hope to show that isn't true."

"Is Professor McGavock your advisor?" Dani asked.

"No, Professor Moreno. I was McGavock's teaching assistant last year, but his expertise is in astrophysics."

"Ben should get his PhD close to the time when I begin my thesis research," Camila said.

"Within a few months," Ben added. "I should be able to get a job here in the Capital District with one of the research companies associated with the university, so I'll still be around to help Camila."

Ryan nodded and addressed Camila. "We should talk about that. At some point, you'll have to think about what you want to do after graduation."

"What do you mean?" Camila asked.

"If you're successful, you'll have a promising concept but not a product. You might take your experience to an existing company, or you could start your own company to exploit what you learn."

"I hadn't thought about that."

Ryan smiled. "It's not something you have to decide on now, but it certainly wouldn't hurt to think about it."

Three large pizzas arrived then, and the conversation ceased while everyone took a slice and began eating. With Ben and Mike leading and Dani not far behind, the pies disappeared quickly.

"Camila?" A tall, slim brunette approached their booth. "That is you."

Camila looked past Ben. "Hi, Gabby. It's me." She glanced around the table. "This is Gabriella Varela. She's in my Schrödinger class." She introduced everyone to Gabby.

Gabby smiled at everyone, ending with a longer gaze at Ben. "Camila, I didn't know you had a boyfriend."

Ben gulped down a mouthful of pizza and smiled up at Gabby. "I'm not. Just friends."

Gabby raised her eyebrows, and the corners of her mouth curled up slightly. Ben had turned away from Camila, so she didn't try to hide her grin. Ben was certainly fast to deny any romantic involvement with her and that was fine.

Ryan might have read the situation as well. "Grab a chair and have some pizza."

Gabby glanced around the table, her mouth twisted as if in thought.

"We were talking about Camila's research and what comes after." Ryan squeezed into the seat next to Mike and motioned Gabby toward the chair he had vacated. Mike squeezed closer to Dani with a grin.

Gabby sat down and looked around the table, glancing first at Ryan with a quizzical look and then at Ben.

Ryan motioned to a waitress. "We're very friendly. Beer?"

"Ah, sure."

"I don't remember seeing you around the Physics Building," Ben said as Ryan signaled for the waitress to bring another glass.

"I'm studying for a PhD in software engineering, not physics."

"Really? Taking a course on the Schrödinger equation?"

Gabby nodded. Camila had an impression it was no longer a group discussion.

"I need quantum physics in my work," Gabby said. "My thesis is on using quantum computing for simulations. I have an idea for a class of algorithms that would be an improvement on current methods."

Dani shook her head. "And I thought Ben and Camila were difficult to understand."

Gabby blushed. "Sorry. Ben asked." She stared into his eyes. "I'm sure he understands."

"I do," Ben said.

"Me too," Camila said, earning a grin from Dani.

Ben ignored Camila's comment. "Simulations have become the most important method for studying complex phenomena. Solving equations in even simple situations like the three body problem isn't possible."

Gabby flashed a smile at Ben. "Exactly."

They continued the gathering with several more rounds of beer and another pizza. It was after ten o'clock when Gabby looked at her phone and gasped.

"I didn't realize it was so late. I should get home."

"Where's home?" Ben asked.

Gabby gave a downtown address almost a mile away.

"I can walk you home," Ben offered.

"Would you? I would feel much safer."

As the two people said their goodbyes and walked away together, Camila and Dani exchanged looks. "Isn't that the opposite direction from where Ben lives?" Dani asked.

"Yes, it is," Camila answered. Both erupted in giggles.

11

IT WAS LATE SEPTEMBER, but the evening was still warm. Sweat formed around the edges of Ben's mask as he and Gabby walked down the hill toward her downtown apartment. The Hudson River, occasionally visible between the buildings, shimmered from moonlight and lights along both banks.

"That was Dani Howell." The mask muffled Gabby's voice. "Isn't she on the hockey team?"

"She was. She's a graduate student now."

"I've wanted to go to a game, but was reluctant to go alone."

Ben hesitated, but took the plunge. "The men's team played tonight. I could go with you to the women's game tomorrow night. I'm sure Dani's old team would appreciate the support."

Gabby's mask stretched a bit. Into a smile, Ben hoped. "I'd love that," she said.

THE GROUP MET AGAIN in the middle of October. They enjoyed each other, and for Ryan, paying for the food was a small price for spending time with his son. Mike wasn't always eager to prioritize time with his father, but Ryan was sure Dani's presence tipped the scales in his favor.

Besides, he liked Camila and the others. They reminded him of another time, when he was a student trying to find his place in life. If Camila did go her own way, he could help create the core group she would need.

That night, they met in the small restaurant in the Student Union, where beer and decent hamburgers were available. They talked a little about Camila's work, but she didn't have that much to report. Their beers arrived, and the conversation paused.

When it resumed, Dani changed the subject. "So, Ben, how come your girlfriend isn't here?"

Ben struggled to find an answer. "How did you know?" he finally asked, red-faced.

"Mike and I still go to the games. You think I haven't noticed you? Gabby is quite the cheerleader. I think I caught you kissing her at least once, too."

"She's not involved in Camila's project," Ben said. "I didn't think I should invite her."

Ryan spread his hands on the table. "Maybe we should get her involved. This is a small, tight group for now, but is she as good at software as she seemed?"

"I think so," Ben answered. "She's helping me with software to analyze the data I'll be collecting."

"She's doing very well in the class I take with her," Camila added.

Ryan lifted his hands and shrugged. "I brought up the idea of Camila forming her own company at our last meeting. Have you considered that, Camila?"

"Some. I don't know what I'll do."

"I think we all enjoy these get togethers. But if you decide to run with any discoveries, having a trusted group of friends will help. When and if that happens, you'll need good software engineers, won't you?" Ryan grinned. "I'm sure Ben wouldn't mind if Gabby joined us."

Ben shrugged, but he was grinning too. "OK, if that's what you want."

T HEY WERE BREAKING UP, and Camila pushed back her chair and stood. As everyone else moved away, Ben put a hand on her arm to hold her back. "Gabby and I are going on a hike tomorrow," he told her. "Would you like to come?"

Camila stared at him, trying to read his intention from his expression but failing. "Wouldn't I be a third wheel?"

Ben grinned. "Usually. But I think Gabby might be a little hesitant about joining us, and having you there to back me up would help."

"Where are you going?"

"One of my favorite places. Severance Mountain, less than two hours away. We can stop and have dinner with my parents on the way back. They haven't met Gabby yet."

"This isn't one of those wilderness hikes into the Adirondacks, is it?"

"No, not at all. I like it when I just want a short hike. It's less than two and a half miles, maybe a two-hour walk. The view is beautiful."

"How will we get there?"

"My dad will lend me his car. He's already told me he approves of Gabby, just from what I've told him about her. You have good walking shoes?"

"Sneakers. Will that do?"

"Sure. Especially if they support your ankles."

That sounded like something more than an easy hike, but Camila suppressed her reluctance. The idea had appeal, too. Ben often referred to the beauty of the area, and she had to admit that a day off from her studies wouldn't hurt. "All right. If Gabby doesn't mind, I guess."

ALTHOUGH AUTUMN HAD STARTED a month before, October temperatures still reached the eighties in the afternoon. They set out for Severance Mountain in the morning when the weather was still comfortable. Ben had gone home to Saratoga the night before and gotten his father's car. When he picked up Camila at her apartment, Gabby was already in the car, and to Camila's relief, her greeting was friendly.

The trip up the Northway took about an hour and a half. In the air-conditioned car, they could take their masks off, and the conversation ran to small talk as Camila and Gabby got to know each other better. As they continued north, the signs of civilization became less frequent, and they spent more time enjoying the Adirondack scenery.

Gabby finally turned to Camila. "So, I guess I'm going to have to ask. Ben was evasive about why he wanted you to join us." She smiled. "I'm assuming it wasn't because he wanted to have a threesome at some seedy motel up here."

Camila chuckled. "Well, forget the seedy motel, and add Ryan Terry to make it a foursome, but yes, actually. We would like you to become part of our group."

"This discussion of your thesis project you mentioned the night I met Ben? Why?"

"It's more social than technical. Nothing much is going to happen until I have my PhD, so you wouldn't be committing to anything. But Ryan thinks I might want to form my own company to develop products from my research. If my project turns into something, we're going to need software engineers. We should have one helping us with that planning, and we think you would be a good fit." She grinned. "Besides, Ben would appreciate it."

Gabby glanced at Ben, but he concentrated on the scenery. Camila watched her face as Gabby's expressions cycled through frowning, scrunching up in thought, and curiosity. Finally, she turned back to Camila.

"All right, so I need more information. Tell me more about this project you're all so enthusiastic about."

Camila told her about Stenhouse and the rest of it. Gabby had questions, and they were still talking when the car left the Northway and doubled back on Route 9 to get to the trailhead, where a small sign marked the dirt road entrance.

"We're going to climb a mountain, but it will be easy," Ben said.

Camila got out of the car and looked around, but all she could see were trees. A few had begun the change to fall colors, but most were still green. The air was already warming, and her doubts returned.

"We're climbing a mountain?" she asked.

"Really more like a hill," Ben said. "It's less than fifteen hundred feet high, and we're already almost nine hundred feet up." He pointed to an elevation sign at the edge of the parking lot. "The view is worth the climb. You'll see."

They found the trail heading west and followed it through tunnels under the Northway. The path was almost level, a pleasant walk through the forest, and Camila's reservations faded. It really was beautiful, and Ben had promised spectacular views. She forgot that splendid views usually required great altitude, and the trail soon rose. At first, the rise was easy, but soon there were steeper parts with rocks and tree roots to avoid. The surrounding forest was inviting, but the trail often required more care and less enjoyment of the scenery.

Camila stumbled once, but Ben grabbed her shoulder and kept her from falling. When Gabby faltered further up the trail, Ben intervened again, but this time, he grabbed Gabby's waist and pulled her close.

"I wasn't going to fall," she protested. But she was smiling and gave Ben a quick kiss as he released her.

They stopped twice to catch their breath after steep sections and arrived at the top with Camila and Gabby breathing a little heavily. Ben didn't seem affected. Camila looked over at Ben and grinned. The view was worth the effort.

She took out her phone and took several pictures. The forest stretched out below them, overlooking a calm lake and dark silhouettes of more peaks beyond. She understood why Ben talked so much about the area.

Ben and Gabby moved a few feet away, and Ben put an arm around Gabby. She stared out, clearly enjoying the view as much as Camila did.

They weren't alone at the peak. Six other people were there, relaxing or taking pictures. One man approached Camila. "First time up here?"

Camila glanced at him. He was probably about her age, with a smile that was his best feature. She pointed over toward Ben and Gabby. "Yes. My friend brought us. He's done a lot of hiking in the area."

She moved closer to the edge of the hill and Ben and Gabby, hoping the curt reply and failure to return the smile would discourage the man. Apparently, it worked, because he didn't follow her. She didn't look back and didn't see him again.

Camila took a few more pictures before they started back down the trail.

B EN SAW THE OBVIOUS enjoyment of his companions, but had to fend off depression. They saw the forest as it was, its beauty new to them, not from his perspective of twenty years hiking in the area. He noticed the increase in dead trees and the signs of damage from temperatures higher than what the trees had evolved from. Pollution had taken its toll, too, although the air was much cleaner than in the cities to the south.

The condition of the trees was not the only sign of problems. Other than birds and a chipmunk, he saw no animals. As a child, he and his father had often come across deer, and once even a black bear at a distance, but most animals had either died or hidden themselves deeper in the mountains, away from the invading humans.

Too many politicians claimed they had solved the problem of global warming, but Ben disagreed. The average global temperature was still rising, although slower than before. Glaciers—those that had survived—were still melting, and species were still becoming extinct.

Camila might have an idea that could help, but there was a problem. Professor Moreno had confided in him that many physicists would watch his research, hoping that he would fail and discredit Stenhouse's results. Few accepted the existence of a previously unknown field, and they could be right. If he failed in his research, hers would be rejected as well.

He knew he should tell her. Maybe he could work up the courage soon, but he didn't want to be the one to puncture that bubble of optimism. He looked at Gabby, still flashing smiles at him occasionally despite her obvious fatigue from the climb. Their relationship was only a few weeks old, but he had a feeling she was the one. Maybe, someday, their children would see the surrounding forests and peaks restored to their past glory.

12

CAMILA AND THE OTHERS finished the hike in the late morning and started back, getting lunch on the way. After that, they continued to Saratoga Springs to visit Ben's parents.

As they approached the house, Ben's father, Bruno Pinto, opened the door and greeted them. Bruno was average in height but moved with the assurance of a physically fit man. That wasn't surprising; Ben often talked about their long hikes and backpacking trips into the Adirondacks. His booming voice provided additional evidence of his vigor.

"Come in, come in." He hugged Ben and turned to examine Camila and Gabby. "From what Ben has said, one of you is his partner and one of you is his girlfriend. You're both so lovely that I have no idea which is which."

Camila blushed. "I'm the partner."

"I guess that makes me the girlfriend," Gabby said.

Ben's mother, Rosalie, joined them. "So get out of the way so they can come in." She pushed Bruno, and he grinned and moved inside the house. The others followed.

"I'm glad you got here early," Rosalie said. "Let's get comfortable and get to know each other."

They sat down in the living room, a large space with wall paintings of beautiful landscapes and two large, mounted fish. Camila was a little surprised there were no animal heads, but guessed that hunting was not a Pinto pastime.

Camila and Gabby spoke about their families and something about their lives, answering questions from Bruno and Rosalie. After a few minutes, the conversation turned to the day's outing.

"Forgive my curiosity," Rosalie said. "I don't understand why you wanted to go up to Severance Mountain with two young ladies."

"I was a little curious about that myself at first," Gabby said. "It turns out your son just wanted Camila along to talk me into joining their little pizza group."

"Did it work?" Ben asked.

Gabby reached over and patted his hand. "We'll see."

"I'm not a businessman like your Ryan," Bruno said, "but it seems to me that ideas like Camila's need large companies with resources. You only have your little group."

"Right now, we're just talking." Camila smiled. "I had an idea, and Ben agreed to help me with it. Ryan found out through my roommate and helped me with a low-interest loan."

"She gets her PhD, so she wins regardless," Ben added. "There may be a company eventually, but not until she finishes. Meanwhile, we plan and maybe find more early contributors."

Gabby frowned. "Early contributors? I liked it better when your dad said I was a girlfriend."

It was Ben's turn to pat Gabby's hand. "I'm confident you can multitask."

Bruno still looked doubtful, but he nodded. "I guess this Ryan guy is the only one who really stands to lose if this doesn't pan out."

"With my degree, finding a job and paying off the loan shouldn't be a serious problem," Camila said.

"But if it pans out, we could really make a difference in fixing the environment." Ben looked at his father with an earnest expression. "If Camila decides not to start her own company, a large company could still buy her research or give her a job to develop it."

"Then we'll hope she's successful." Bruno smiled at Camila, and she smiled back.

G ABBY JOINED THE GROUP at the next meeting a month later, just before the long Thanksgiving weekend. Now six strong, they needed a table instead of a booth when they returned to the pizza restaurant where they first gathered.

Gabby glanced at Ryan Terry as they ate. He was the odd person in the group. Everyone else was in their twenties, but their host was at least twenty years older. Probably more, since his son, Mike, was out of school and working for his father. But the elder Terry looked comfortable with the younger members. He appeared to enjoy the byplay of the others and occasionally joined in. He seemed proud of Mike, too, and Gabby decided Ryan must be a wonderful father.

When Gabby had hinted that she might not go home for Thanksgiving, Ben invited her to his family's dinner. She accepted quickly. Her own father was nothing like Ryan or Ben's father, and she had enjoyed meeting the Pintos after the hike.

Ben had invited her into the group, and Camila convinced her to accept. Ben was clearly attracted to her, and she felt the same, although he was definitely a teek. Or maybe that was part of his desirability; she was somewhat enamored by technology herself, although her preference was software.

According to Ben, the invitation had been Ryan's idea. Ryan wanted to pull her into this project that had brought the group together, but she still wasn't sure she wanted that. A future with Ben was possible, but she was wary about an effort unlikely to succeed. In the twenty-first century, large, well-financed companies made significant advances in technology, not a few students eating pizza.

Of course, she wasn't really committing to anything beyond socializing. Her older brother had already started his own successful software company and promised her a place in it if she wanted it.

Camila was looking at her strangely, either out of concern or curiosity, and Gabby realized she had drifted away from the conversation. She placed a hand over Ben's, and he grinned at her. That seemed to satisfy Camila.

RYAN SWALLOWED A BITE of pizza before addressing the group. "The school plans to build a new physics facility. I've talked some people I know into donating as well, and the college was able to inspire a couple more donations."

Ben nodded. "They're calling it the Energy Innovations Laboratory."

"Professor Moreno talked to me about it," Camila said. "She has a lot of input into its design. I'll have a laboratory there."

Ryan had loaned Camila significant money to finance her doctoral program, and promised an even larger donation to the college. He didn't know what the end result would be, but, as he sat at the table thinking about the group, he realized it was probably the best money he had ever spent.

He hadn't been around much when Mike was young, too busy making his fortune to spend much time with a small boy. Later, when it sank in how much he was losing, it was almost too late. His wife, Ana, divorced him after fifteen years of marriage because he didn't spend enough time with her either. Mike had been twelve and Ana got custody and moved to Arizona, where she had grown up. Ryan had visitation rights but rarely used them.

When it came time for college, Ryan didn't complain about paying for it. Mike was his son, he could afford it, and his responsibility seemed clear. Mike surprised him by choosing Albany State rather than a college in the southwest, but he soon understood that Mike was curious about his absent father.

That had been awkward. Ryan had no idea how to bond with his son, but he tried, even hiring Mike to work in his investment firm. Now, looking at the faces around the table, he knew he had finally succeeded. Camila's research idea, whether it became anything, had created this circle of people where he felt comfortable with his son at last. His investment up to then was insignificant compared to what he would gladly sacrifice to keep them together.

13

SENATOR LEO CORREIA STARED at the opposition party members across the aisle of the Senate chambers. With a procedural vote coming up on the Newman-Fredrickson Privacy Act scheduled for that afternoon, he assumed they were conspiring about how to convince more senators to vote to consider the version already passed by the House.

If the motion passed, the Senate could begin debate on the bill. Correia planned to have his say, but he doubted his fellow senators would listen. Not enough of them, anyway. The bill had passed easily in the House, supported by Congress members who thought making it harder to reveal what they did was a good thing.

A massive campaign in the bill's favor had gotten much of the public behind it, as well. Supporters willing to contribute to the campaign, especially rich supporters, had been easy to find.

Who could criticize a law designed to keep the government out of your business? That was the message drummed into everyone unrelentingly from every media source. His party fought back, but the last election had eroded the power they had enjoyed before. No one wanted to hear about how the bill would affect law enforcement by making it almost impossible to investigate anything without first justifying a warrant.

The less-desirable elements of society were happy. Background checks wouldn't expose their revolting secrets, allowing them to engage in their unnatural activities without fear of shame or retribution.

Someone touched his shoulder, and he turned to see Senator Valerie Peterson, a freshman member of his party. He could see concern on her dark face and smiled.

"Cheer up, Val," he said. "We'll have our chance to fight this."

"But we'll fail," she answered. "Support has been building for stringent privacy laws for over twenty years. The only surprise is that it didn't happen years ago. This will pass before the end of the session."

"True, but no one will be able to find out about it without a warrant."

The weak joke at least got a hint of a smile, and Correia smiled back.

"I assume you will speak against it," Peterson said.

"Of course. They'll have to drag me off the floor. My people say they've got five of ours that will vote in favor, though. That probably makes sixty-seven in favor, assuming Baker doesn't go against the tide."

"The climate bill will be next."

Correia nodded. "I think we'll beat that. A lot of people on the other side realize that we've already solved the problem, assuming there was a problem. We just have to wait for everything to settle back down to normal. These things don't change overnight."

"We could try to change a few votes with some key dinner invitations tonight."

Correia shook his head. "I promised my wife I would be home. Good luck, though. If you had a husband to go home to, you could avoid that exercise in futility too."

"If I had a husband, I probably wouldn't have become a senator. Not many Black men want a wife with more power than he has."

"He has to be Black?" Correia grinned.

"Not many white men either." She gave him a gentle push. "Go take your seat. I think they're ready to vote."

Both senators returned to their places. An hour later, the president pro tempore announced the results. By a vote of sixty-seven to thirty-three, the senate resolved to begin debate on the Newman-Fredrickson Privacy Act. At least for the procedural vote, Senator Baker voted with the opposing party.

C ORREIA HAD CALLED PRISCILLA before he left the Capitol, so his wife was not surprised when he arrived home in time for dinner. It didn't happen often, and she had promised something a little fancier than usual. She was seasoning two pork chops when he walked into the kitchen and pulled her into a hug from behind.

She leaned back and rested her head against his shoulder. "Dinner will be in about thirty minutes. You could probably use the time to relax a bit first."

Correia tightened his grip on her waist. "I am relaxing. I passed up an arm-twisting dinner with Val for this."

"It's not Val's arm you need to twist, is it?"

"No. She suggested we work on someone who hasn't quite decided yet. It wouldn't do any good, though, and I would much rather be here." He kissed her neck.

Priscilla chuckled. "That dinner estimate starts when you let go so that I can work on it."

Correia kissed her neck again and let his arms down. "All right." He looked around the counter at the cut vegetables, spices, and pork chops laid out for processing. Behind him on the stove, a pot of water was heating, and a box of pasta sat nearby. He smiled. His schedule hadn't allowed for a good home-cooked meal for more than a week.

"Did you vote on that bill you were worried about?" Priscilla asked.

"Yes, but it was only a procedural vote to allow debate. It passed, but I expect a lengthy debate before we vote on the bill itself."

"What happened to that amendment you were going to try? You said that would slow passage."

"Before they voted on opening debate, the other side pushed through a resolution against changes until we voted on the House version. If that vote fails, which it won't, we can take up changes needed to make it acceptable to the Senate."

Priscilla turned and checked the pot of water, but it wasn't boiling yet. "There have been so many stories about the government spying on its citizens. Many people think we need this law. They say it will make it harder for those annoying ad calls, too."

"Some of that is exaggerated. For a minor improvement in those areas, we'll be making law enforcement impossible."

That's probably an exaggeration, too. But she didn't say that.

"The liberals who are pushing it don't care about that," Correia continued. "They'll trot out a bunch of papers from liberal scientists spouting so-called 'facts' about the damage to democracy that government surveillance does. What's a little unsolved crime compared to their precious right to privacy? Most of them are probably gay."

"If the government didn't use surveillance to violate their privacy, they wouldn't care as much." Priscilla frowned, knowing that she should have kept that thought to herself.

Correia shook his head and left the kitchen. Priscilla breathed a little easier, sure that he would forget the conversation by the time they ate dinner. Sometimes they argued over remarks like that, but not tonight, and her husband didn't let their disagreements cause bad feelings for long.

She looked through the entryway, but he was gone, probably on the computer in his office until dinner. She shouldn't have made that comment, although everyone knew government agencies were spying on citizens and had been for decades. Advances in artificial intelligence over the last thirty years gave those agencies more ability to pry into private lives than seemed right to her. Her husband would say that the innocent had nothing to fear, but it seemed to her that he refuted that by his comment about homosexuals.

Leo's best friend in the Senate was Valerie Peterson. She wasn't married, and Priscilla thought it would be ironic if Val was single because she was gay. Leo wouldn't believe that, and Priscilla didn't really either. But Priscilla also knew that Val would have little chance of holding her senate seat if she were. Even Leo would turn against her.

Their generation had grown up when many people discriminated against blacks. Not Leo, one of the things she loved about him. Although it didn't happen often, he showed a different attitude toward homosexuals.

The water was boiling now. Priscilla shook her head and returned to her cooking.

14

B EN COULD TELL BY Professor Moreno's expression that something had happened, but he couldn't tell whether the news was good or bad. He took the seat across from her and looked at her expectantly.

"We have an arrangement with the Space Force for the use of Collins Station for your research," the teacher said. "There's been a hitch, though."

"What's the problem?"

"They aren't allowing remote management of your experiments. You've got to go to Collins and perform them onsite."

Ben stared at her, speechless. Collins Station was an orbiting laboratory, managed by the United States Space Force. Stenhouse had worked in a laboratory on Earth, and his critics suggested that its gravity affected his results, leading to a false conclusion. Ben had proposed duplicating Stenhouse's experiment in orbit, where gravity wouldn't be a factor. He hadn't planned on spending time in space.

"Getting you a berth on the station won't be a problem. You'll have to stay up there for six months at a time, though, with a three-month recovery period between," the professor told him. "That's the schedule according to regulations for non-military personnel."

"Did they say why I couldn't work remotely?"

"The scientist I talked to was concerned about security. It would be too easy to intercept transmissions to the station. It would have required time for station personnel to set up your apparatus, and that seems to be an issue as well."

"My research isn't classified."

"Haven't you been keeping up on the news? The entire country is paranoid about their data getting stolen. There's a bill being considered now that would really clamp down on the storage of all kinds of information."

Ben's mind felt numb, but he got out one more question. "When do I leave?"

"Not until early October. That's when the next shuttle goes to the station." Moreno smiled. "Some people pay a lot of money for an opportunity like this."

It was Thursday, and Ben would have to tell everyone at their Friday meeting, but one person had to be told before that. Gabby had a test the next day, and they hadn't planned on seeing each other that night, but he called her just before the hour to avoid the possibility of catching her in class.

"I need to talk to you," he told her when she answered. "Are you on campus?"

"I'm heading to my last class of the day," Gabby answered. "Is something wrong?"

"Meet me at the student union after your class. I don't want to do this over the phone."

He could see the worry on her phone image and tried to smile. "Don't worry. I'm not breaking up with you."

The attempt at levity failed. But she agreed to meet, and they broke the connection.

A BALCONY ON THE second floor of the student union overlooked the lobby, and a couple could sit comfortably with little noise to prevent a quiet conversation. Ben told Gabby what Professor Moreno said, and he could see it disturbed her, but she stayed calm. She asked when and for how long, but then fell silent.

Ben put an arm around her shoulders, and she pushed in closer.

"I know about the security bill," she told him. "It's an attempt to protect illegal data access by placing restrictions on connections between computers. The social media companies oppose it, of course, and even law enforcement is trying to weaken it."

"Law enforcement?"

"Sure. It could mean needing a warrant for data that they now have easy access to. Vehicle licenses, for example."

Ben tapped her nose with one finger. "And how do you know so much about it?"

She wrinkled her nose and swatted his hand away. "I'm going to be a software engineer. This will affect everything I do. I've already been told that the school is putting together a course that I'll probably take next semester if the bill passes."

"But it hasn't passed yet."

Gabby shrugged. "No, but the Department of Defense isn't waiting. Some other government agencies with endangered data are going forward with the policy on their own as well."

"So I'm stuck with it."

Gabby kissed him. "We're stuck with it."

THE USUAL FRIDAY GATHERING was subdued. Ryan tried to lighten the mood but met only awkward nods and silence. Dani tried next.

"We could all go to the hockey games together. Have any of you been since I graduated?" She reached across the table and patted Gabby's hand. Dani had been a year ahead of Camila and Gabby and was now working for Ryan.

"Ben and I go to them sometimes," Gabby said, with no visible enthusiasm.

Mike was more excited. "That sounds like a great idea. It's not the same without Dani intimidating the other teams, but it'll still be fun."

Ryan nodded. "I could probably make it occasionally, too." He looked at Camila. "How about it, Cam? A group event?"

Camila's smile didn't have the usual dazzle. "Sure."

"Sounds like you're all going to party after I leave," Ben said.

Gabby grabbed his hand. "We're going to need something to distract us away from missing you, babe."

"Well, some of us will," Mike said, eliciting the first chuckles of the meeting, weak though they were.

FIVE DAYS LATER, BEN received a call from Space Force Captain Foster.

"We will take you up to Collins Station on October 7th," the officer told him.

"So soon? That's less than a month from now."

"Please be in Orlando the day before. Notify my office of your flight number and we will pick you up at the airport and bring you to the Kennedy Space Center by helicopter."

"All right."

The captain frowned. "Try to sound more enthusiastic, Mr. Pinto. Your government is providing you with a valuable opportunity so that you can pursue your research."

"Yes, sir. Sorry. I had expected to run my experiments from the surface. It came as a shock when I was told I would have to go to the space station."

Foster's expression softened. "I understand. You'll appreciate the opportunity more when you get to the station. It will be the experience of your life."

"Have you been there?"

"This will be my third trip. I'm very much looking forward to it. And don't worry. We haven't lost a civilian yet."

"I know. I wasn't worried."

Foster smiled and nodded. "Ah, leaving a girl behind?"

Ben shrugged.

"She'll be there when you return. I've been married fifteen years, and, trust me, they're always overjoyed to see you again."

15

CORREIA MET SENATOR VALERIE Peterson for a quick lunch in the Senate cafeteria. The current session had been scheduled for a one-week recess, but the majority leader canceled that to force a vote on the Newman-Fredrickson Privacy Act. The bill had passed that morning.

"This will be an issue next year when crime rates skyrocket," Correia said. "They'll have to explain all this to their voters."

"I'm glad we won't be up for reelection," Peterson responded.

"Something has to be done. We can't even access voter registration records, and that's going to hamper campaign strategy for both sides. They'll be trying to find a fix, and maybe we can get the jump on them."

"What do you have in mind?"

"Draft a quick bill to alleviate the effects. Expand the use of voluntary information release, for example. We might even get Baker as a cosponsor."

"That would work in our favor." Peterson paused. "I could draft something up."

"Thank you. I'll try to get the ear of a few others on the idea so we can push a bill through as quickly as possible."

"Pressure from the public would help."

Correia nodded. "I'll schedule an appearance on a couple of talk shows next week to publicize it. You should too."

Correia's phone buzzed, a signal that usually meant significant news. "Just a second. Let me check this." He touched the phone's screen, and a news report began playing. He listened for a minute before looking at Peterson.

"The stock market has already started responding, and not in a good way. All the indexes are down, most by a couple percent."

"Because of the privacy act?"

Correia nodded. "A lot of companies depend on information that will be harder to get now." He paused before smiling. "This could help us."

"Most people think the stock market is for the rich. Getting the privacy they think they need might be seen as an acceptable tradeoff."

Correia snorted. "That may be truer of your constituency than mine."

Peterson straightened in her seat and scowled at Correia. "You think Black people don't invest in the stock market?"

Damn, that's not what I meant. "This has nothing to do with color. I don't mean to sound like a bigot, but Texas is less affluent than New York."

Peterson's face relaxed a little. "Not that much, but I guess I'm a little on edge right now. We opposed this bill, but some of the blame will fall on us, too."

"Let's get those talk show appearances."

C ORREIA HAD A SPOT on the talk show Washington Views the following Tuesday. The show tended toward conservative views but was moderate enough to be considered bipartisan. Its host, Kendra McDermott, was a stunning brunette known for her sharp insights and pointed questions. Extremists had learned to avoid her, but more moderate politicians coveted a spot on the show. Correia had always gotten along well with her.

"You voted against the Newman-Fredrickson Privacy Act," Kendra said after the usual introduction.

"I did," Senator Correia answered. "Along with many in my party. We felt it was an extreme solution to the problem of privacy, and we've already seen the consequences."

"You're referring to the stock market."

"Yes, of course. The plummeting values of securities have erased a significant fraction of the wealth of this nation. That affects many more people than the rich. Middle class retirement plans, jobs, and our ability to communicate are already being weakened by the crash. Even I didn't foresee how catastrophic the immediate effects would be."

"You were more concerned about the long-term effects."

"Law enforcement agencies across the board opposed this bill. We must protect the privacy of our citizens, but their safety is important too, and this will cripple the ability of the police to solve crimes. The law is just too extreme."

"What can be done about it?"

Correia smiled. "Senator Peterson and I are already working on a bill that will ease some of the restrictions and make it possible for the police to get information about legitimate suspects in investigations. As it stands, the Newman-Fredrickson act doesn't even allow people to volunteer their information to be accessed across computers without stringent warrant constraints."

"Your bill would let people opt in to law enforcement access to their information. That will help, but criminals won't do that."

"You're right, of course. But we also intend to loosen the restrictions on integrating computer access. Because of this bill, the police can't even get DMV information without a warrant. Our bill would preserve the distinction between government computers and commercial or private computers, but allow most communication from one government computer to another. Commercial entities would still have strict protocols about enforcing user privacy by default."

Kendra had a few more questions that Correia had no problem answering. The interview was going well when Kendra surprised him by moving sideways from the intended topic.

"To get back to the economic effects of this bill, I have an example of a setback in your own state, Senator. An engineering college in New York has halted construction on a new research facility because funding has become a problem."

"I was not aware of that. What facility?"

"It's called the Energy Innovation Laboratory. The school was forced to put the project on hold."

"That's unfortunate, but I'm sure there are many projects in my state and in the rest of the country that suddenly have funding problems. Senator Peterson and I hope to address them with our legislation."

"According to the school, the facility is being built to explore new green energy options to address the global warming problem."

Correia smiled. "That problem has been solved. The planet just needs some time to settle down. Alternative energy sources would be nice, but we're looking at more serious issues than global warming."

"Not everyone agrees with your description of the state of the climate. Including most scientists."

"And they are entitled to their opinion. The fact is that programs put in place more than a decade ago have reduced carbon emissions to where atmospheric carbon dioxide levels have almost stabilized. It takes time for carbon dioxide to leave the atmosphere, though, and some extremists can't wait. My colleagues and I have the duty to prioritize the most serious problems. My position on this has been clear."

Kendra nodded. "Thank you for coming, Senator. I'm afraid that's all the time we have for this discussion."

"You're welcome, Kendra. Thank you for having me."

16

WHEN PRESIDENT KALEB WATTS called Ryan and asked him to come in to talk, Ryan knew it wouldn't be good news. The stock market wasn't recovering, and the best that market watchers could say was that manufacturing company securities were stabilizing. Technology companies were not.

Watts' expression as Ryan took a seat did nothing to relieve his trepidation. He suffered through the usual pleasantries before the university president addressed the reason for the meeting.

"You're probably not surprised that several promised endowments for the Energy Innovations Laboratory have been delayed or canceled," Watts said. "Construction will have to be postponed if we can't replace those funds."

Ryan nodded. "I'll probably have to spread my own endowment out over time. I understand the problem."

"We're trying to maintain all the programs here, not just the new building. Possibly there are other sources of money, but the university has the resources to pursue them for only the highest priorities."

"You could pace the construction to what's available," Ryan said. He knew that was unlikely to be practical, but he was feeling desperate.

"I believe some of the promises came from associates of yours. The university can't dedicate our fund raising to such a speculative project, but perhaps you can find alternative sources."

Speculative meaning the Stenhouse Field. "And if I can't, how will that affect Camila Lopez's doctoral research?"

"We don't know what she'll need yet. I hope we can find a place in an existing facility, but I've been told that might not be possible. If her requirements are too expensive, she may have to change her subject."

Ryan could do nothing about the situation immediately. "Thank you for informing me."

"She won't begin her research for some time. Perhaps the situation will change."

"I'll talk to Ms. Lopez. I'm sure she will adapt to the new circumstances."

"Y‌OU KNOW WHAT HAPPENED on the stock market this week," Ryan said to everyone at their next meeting.

"The crash?" Dani nodded. "It's because of that bill, right?"

"Yes. Social media companies only exist because of access to personal information that will now be illegal for them to hold." Ryan looked around the table, dreading what he had to tell them. "It's not just social media, either. Many companies depend on the exchange of information, and limits on what they can keep will hurt their business too. Especially in getting new customers."

"Why didn't they realize that was going to happen before the bill passed?" Camila asked.

"There was a lot of pressure to fix problems in abusing information," Ryan answered. "Even the government has been doing it for decades. A lot of politicians are up for reelection next month, and they probably hoped the hammer would drop after they were reelected."

"Dad has lost a lot of money already and will probably lose more next week when the markets reopen," Mike said.

Ryan nodded. "Not just me. People who promised endowments for the energy facility have backed out. The school has put off its construction indefinitely."

He looked at Camila and could see the worry lines on her face. "Camila, the money for your PhD has already been paid in advance, so this doesn't affect your degree."

"Except that it might affect my thesis research," Camila said. "Where would I do it if they don't finish the Energy Innovation Building?"

"That could be a problem," Ryan said. "The school hopes to support you in an existing facility. It may make planning for what you will need a bigger priority, though. That could make a difference."

"ARE YOU GOING TO be all right?" Mike sipped on his tumbler of bourbon and looked at his father. He and Ryan had gone back to Ryan's house after the meeting, and now they relaxed in the study.

"I think so," Ryan said. "I have a core of very conservative investments that pretty much guarantee I won't go broke. The house is paid for, so I'm in no danger of losing that."

"So this doesn't affect you?"

Ryan grimaced. "Oh, it affects me. That core is only about a third of my total investments, and the rest got hit hard. That's money I wanted to have if Camila started her own company. It would be a promising investment."

"You didn't say that at dinner."

"There's no point in worrying the group and I didn't want to assume anything about what she will decide."

Mike nodded. The thought intruded that the possibility of a future goldmine might be the reason his father was interested in Camila's education, but he didn't think that was true.

"I shouldn't need the money until Camila graduates, which gives me a few years to recover," Ryan continued. "I think the government will do something to reduce the damage, so I'm optimistic."

"You don't look optimistic."

"I expect some rough times for a year or so. I'm optimistic about the future beyond that."

RYAN HAD A MEETING in California the next week and returned late Friday night. There was a general feeling that the group was due for another meeting, so on Saturday night, they all went to the hockey game and out for pizza afterwards. Many people had the same idea and there were no tables available, but they found an empty booth and crowded in with Dani, Camila, and Gabby on one side and Mike and Ryan on the other.

"Have you heard from Ben?" Ryan asked Gabby.

"Two days ago," Gabby answered. "He's all settled in and raving about the view from space." She frowned. "It's been less than a month, but he seems to enjoy it."

"The novelty will wear off soon," Dani said.

"Sure, maybe then he'll miss me and we'll both be miserable."

A server came to the booth, and they ordered beer and two pizzas.

"How about your social life, Camila?" Mike asked.

Dani snorted. "She has no social life. These meetings are the only time she gets away from her computer."

"I've got a couple of tough courses this semester," Camila protested.

"Well, you're not studying now," Dani said. "Greg is over there with a couple of friends." She waved toward a table across the room. "A night out occasionally wouldn't hurt you."

Camila wrinkled her nose. "Not interested."

Dani shrugged. "There are other guys around."

"Still not interested."

The server brought their beers. Ryan ignored the byplay about Greg and turned the conversation back to Camila's studies. "I'm sure you'll do fine on those courses," he said. "You don't want to burn out."

Camila looked at him, startled. "I won't. Honestly, I enjoy it. That's why I got into physics."

"Stereotypical teek," Dani said. But she was smiling.

"That's me." Camila bumped shoulders with Dani and smiled back.

Later, after the pizzas arrived, the conversation slowed. Camila was reaching for another slice when Dani nudged her.

"Greg is coming over," Dani said.

Camila frowned. "Don't look his way, and maybe he'll take the hint."

It didn't work. Greg stopped by their booth and leered at them. "So where's your boyfriend, Cammy?"

When Camila didn't answer, Greg smiled and leaned over the table. "I heard he left you for someone else."

"If you're referring to Ben, he was never Camila's boyfriend," Dani said when Camila remained silent. "And he didn't leave. He's on Collins Station doing graduate research and will be back soon."

"Uh, huh. Maybe you should find someone else then, Cammy. Ben's not your boyfriend and you won't go out with me. If you're not careful, people might think you're gay."

Camila flushed, and Greg chuckled. "If Ben's not interested in you, maybe he's gay, too."

Gabby laughed. "Oh, I assure you, he's not."

Greg looked over at Gabby. "So you're the one that took Ben from Cammy."

"I didn't take him from anyone," Gabby said.

Camila scowled and deliberately turned away from Greg. "You're disgusting, Greg. Get it through your thick skull that I want nothing to do with you."

"Then you must be gay."

Ryan had been quiet until then. Mike was on the inside, against the wall, and Ryan was on the aisle side. He stood. "I think you should take Camila's advice and leave. Take your adolescent crap somewhere else."

Greg turned his attention to Ryan. "Who's going to make me?"

"Wow, so clever!" Ryan lifted his phone. "I have the police on speed dial. Shall I call them?"

"I didn't do anything."

"You just threatened me, in front of witnesses. Think you can convince them otherwise?"

Ryan tried to maintain a cool image as Greg's face twisted with anger. He could be in trouble if Greg got physical, but Ryan hoped a confident appearance would prevail. It was a relief to see the anger give way to uncertainty before Greg broke eye contact.

"Fine," Greg said. He looked at Camila. "I don't know why I should talk to you losers anyway." He stalked out of the restaurant.

"And you wonder why I don't date," Camila muttered. She took the pizza she had been reaching for and bit into it, eyes on the plate in front of her.

Ryan sat down and looked at Mike, but Mike only grinned at him and gave him a thumbs up. Ryan sighed and picked up his pizza slice.

17

A T THE NEXT TWO meetings, Ryan tried hard to be optimistic when he talked to the others, especially Camila. Except perhaps for her grades, having a place for her research was Camila's primary concern.

He saw her discouragement in the absence of her smile and her lack of involvement in the conversations of the others. Unsuccessful in convincing her something would be available when she was ready, he sought support from the government.

The Senate ended its session the following week, and Senator Correia returned to his home in Rochester. Ryan had enough influence to get an appointment and took a flight from Albany.

An icy wind off Lake Ontario chilled him as he boarded a taxi at the airport and told it to take him to Correia's office. The disapproving frown on the assistant greeting him at his destination did nothing to improve his mood, but he squared his shoulders and forced a smile as he was ushered into the senator's office.

Senator Correia shook hands with him and motioned him into a seat with a smile that didn't reach his eyes. He glanced at the computer monitor on his desk before turning to Ryan. "So, what can I do for you, Mr. Terry?"

"Are you familiar with the Energy Innovations Laboratory proposal?" Ryan asked.

"I've heard about it."

"The college depends on endowments to fund its construction, but that has become difficult in this economy. Planning is complete, but they don't have the funds to build."

Correia frowned, and creases formed above his eyes. "You work for the college?"

"Not directly, Senator. I've made an endowment, so I have an interest in its success."

"Why isn't the university lobbying me on this project?"

"Frankly, because they don't have the resources to even ask for the resources."

Correia's eyes narrowed. "Don't try to play me, Terry. What you really mean is that it's not high enough on their priority list."

Ryan cursed himself silently. *I'm blowing this.* "I'm not sure I see the difference. Yes, they have other needs besides this project. That doesn't change the fact that it's critical to this country to develop better sources of power."

"It would be welcome, but not critical." Correia's eyebrows lifted.

"This country's need for clean, inexpensive energy grows every year. We can't get it by burning fossil fuels, and other known sources can only take us so far. Alternative sources are needed, and this facility can develop them."

"Other government and commercial facilities are working on power generation possibilities. Why do we need one more?"

Correia's reaction was a disappointment. Ryan had hoped to interest him on the project's merits without bringing in his personal connection, but it hadn't worked.

"A graduate student has an idea and is pursuing it for her doctoral thesis. I'm helping her financially, and the facility would support her research."

Correia nodded and frowned at Ryan. "So you want the government to spend its citizens' money helping you get rich."

"I have contributed an endowment for this project, not invested in it. Others were also about to make endowments, but the crash triggered by the Newman-Fredrickson Privacy Act has forced them to reconsider. Projects like this could help the economy recover."

But he could see that Senator Correia wasn't buying it. The politician's eyes flickered at the mention of the privacy act, and Ryan knew he had voted against it, but that wasn't enough.

"I'm sorry, but can't help you," Correia said. "And I have work to do. My secretary will show you out."

R YAN HAD HOPED THAT being one of Correia's constituents, lobbying for a project in the senator's state, would be in his favor. On his return to Albany, he spent a day making calls to less likely prospects. He talked to a deputy director in the Department of Energy, and that woman listened to him for almost five minutes before telling him she could do nothing for him. That was better than the reception he received from others he tried to talk to.

The next day was a Friday. Ryan usually looked forward to their meetings, but that night was an exception.

"The money to continue construction just isn't available right now," Ryan told them. "We can hope that things will settle down in a few months. This isn't the end."

"When things do settle down, there will be other projects wanting to get moving again," Dani said. "It could be more than a few months."

Ryan sighed. "Yes, it could. But we weren't going to really move forward until Camila completes her classwork. The facility could be finished in six months or less once the money is available, so we have time."

"I met with Professor Moreno yesterday," Camila said. "She also thinks I should drop fusion from my plans. If our speculations about fission are correct, that has a better chance of being successful at a reasonable cost. Only if the glow Stenhouse reported really is the field tearing nuclei apart, though. We know it affects gravity, and the glow might mean it also affects the strong force."

There were spiritless nods. Then they waited for their food in silence.

Camila groaned, and Ryan looked up to see Greg Mora approaching their table. He was on the far side of the booth with Camila and Gabby, and Greg had to pass Mike and Dani as he reached the table. Ryan started to rise, but Greg suddenly stumbled forward.

He grabbed the edge of the booth to prevent himself from falling and glared at Dani. "You tripped me."

"Sorry," Dani said. "I didn't see you coming."

If Greg had intended to say anything to Camila, he forgot about it. He cast another angry look at Dani and stalked off, swearing.

Ryan looked at Dani with raised eyebrows, and Dani grinned. "That worked on the hockey rink, too."

The incident raised their spirits, and when their pizzas arrived a few minutes later, they attacked them with gusto. At least now they were silent as they concentrated on eating and not from depression.

Everyone seemed to be in a better mood leaving than they had been coming in.

18

B EN'S SHUTTLE FROM COLLINS Station landed in Houston. It was a Friday night, and he couldn't get a flight to New York until the next day. Ryan had an automated van that easily held everyone with room to spare for Ben, so the entire group went to the airport in Albany on Saturday to meet him.

He came down the escalator without baggage and wobbled a bit as he reached the bottom. A flight attendant walked with him, carrying two bags that were probably his.

Ryan started forward, but Gabby reached him first, pulling him into a tight embrace. The flight attendant looked amused, and Ryan approached her and pointed to Ben. She smiled and handed him the two bags.

"He's a little unsteady on his feet," she said.

"Thanks for helping him. He'll be OK."

The flight attendant nodded and walked away. Gabby loosened her hold on Ben, and he almost fell over, but he grinned as he regained his balance.

"I guess I was missed."

"Some of us more than others," Camila said, beaming. "But welcome back to Earth."

Ryan wanted to ask him how his research was progressing, but hesitated. Ben had communicated little about that from Collins, and didn't look capable of an extended conversation now. Being in orbit for six months had taken a toll.

They trudged through the terminal to Ryan's van. Ben's face was drawn, and Mike had to help him as Gabby directed him into the back seat. Gabby hovered over him, looking worried, but Ben smiled.

"I'm all right." He kissed Gabby quickly. "It's just moving in normal gravity after six months being mostly weightless. It's going to take me a while to get used to it again. An hour a day in the station centrifuge is no substitute for real gravity."

"Centrifuge?" Dani questioned.

Ben nodded. "The station has one section that rotates to simulate weight. We had to take turns using it, but it was better than what early astronauts had on the ISS and other orbital stations."

At Ryan's command, the van pulled away from the curb. Ben settled back into his seat with a sigh as Gabby snuggled up to him. He put an arm around her, but from the look on his face, Ryan knew it took an effort.

"Up for a celebration meal, or should we take you straight home?" Ryan asked.

"I'm weak, but otherwise OK. Hungry, actually. The food on Collins isn't bad, but I could go for a good hamburger."

Ryan gave instructions to the van, and ten minutes later, they were seated in a restaurant booth near the airport. Once they had given their orders, the catching up began.

Camila voiced the question Ryan had been reluctant to bring up. "Tell us how your research is going. Were you able to duplicate Doctor Stenhouse's results?"

Ben's face turned solemn. "I think we all realize that my thesis project was approved in part because some scientists hoped I would prove the Stenhouse Field doesn't exist. Professor Moreno was the only one who thought I might confirm his discovery. I have the answer now."

Ryan frowned. Ben didn't look happy about it. He glanced at Camila, and she looked stricken. *Has this all been for nothing?*

Suddenly, Ben's expression morphed into a big grin. "It works! Stenhouse's generator produced the field in zero gravity too. It's not some strange gravitational effect." His smile faded. "I still have work to do, though, so I'll be going back in three months."

Gabby pulled his free arm around her. He kissed her forehead, and she responded with a weak grin.

The noisy mix of congratulations and curses from the others over Ben's act drew attention from everyone in the restaurant. Ryan let it go on for a minute before he tried to quiet them down. A curious but smiling waiter came with their orders, after which the conversation was more subdued.

"Tell us about your time on Collins Station," Mike said. "Are the views as fantastic as they claim?"

"Better. Much better," Ben answered. "I was too busy to spend too much time staring out the windows, but I did it as often as I could."

"Earth from space?" Gabby asked.

"That was the most incredible, but not all of it. You can't really see the stars when the sun is visible because of the glare, but when the sun is behind the bulk of the station or set behind Earth, the stars come out. We can't see the stars here in the city, but you've all seen the difference when you get away from the lights and pollution."

Ben paused and shook his head. "You wouldn't believe how much better you see everything, even compared to that. Not just the stars. The moon is incredibly sharp, and I could see the four Jovian moons and the rings of Saturn without a telescope."

Everyone wanted details, but Ben put them off. "The food is getting cold. You can listen to me chatter about it any time."

Ryan turned his gaze toward his food, but stopped at the look on Camila's face. Everyone else was excited, but her expression was different. It was hard to interpret it, but her thoughts were clearly somewhere else. She shook her head slightly as if waking up and picked up a French fry.

Ryan had a sudden insight. "Thinking about those stars?" he asked Camila.

Camila smiled. "You got me. Ben's description reminded me of nights in Texas, away from the lights of Houston. I think my interest in space travel started with those nights, and my father's stories about the Houston Space Center."

Ben must have been hungry, because his attention was completely on his plate. He finished his hamburger before anyone and talked more for a while. After that, the conversation turned to business.

"I've been trying to find money for the school to build the new facility," Ryan told them. "The school says they'll find space, and Ben's success will encourage them, but I'm not sure we can count on that until Camila can tell them what she needs."

"I've been designing a device to investigate the glow effect," Camila said. "For that, I'll need machining support and other help, but Professor Moreno says the school can supply that."

Ben nodded. "They did for me. There's a mechanical engineering student there, Martina Patel, who did great work."

Twenty minutes later, Ryan leaned back in his seat. Mike and Dani had cleaned their plates, but Camila and Gabby were still picking at their French fries. He still had a few French fries left, but he was too old to eat like a college student.

"I've been thinking," Mike said.

"That's what I pay you for," Ryan replied.

"Ha ha. Anyway, I saw a story this morning about some media companies contracting with foreign companies to gather information they can't in this country."

"It won't work," Ryan said. "Not for long, anyway."

"I know. The government will come down on them. But that got me thinking. If we can't get support here, what about other countries? They would benefit from Camila's success, too."

"The crash here has affected other countries, too. Not as drastically, but I doubt we could find anyone willing to endow an American college for such a long shot. Ben's research will help, but I don't know if anyone will understand the application potential."

Mike looked at him with his head tilted to one side and shrugged. "Not an endowment to the college, but what about an investment?"

The idea took Ryan by surprise, but it had possibilities. "It would be even more important to show that the glow Camila talks about has real potential. And if we can get investors, that means giving up a piece of any company Camila starts."

"How much?" Camila asked. "If we're unsuccessful, we won't lose anything more than what we've already put into it. Won't we need investors eventually anyway?"

Ryan nodded. "True. We could try. We might even find investors here." He paused for a few seconds and looked at Camila. "That would mean deciding about your future now."

"I couldn't do it alone. I don't have any illusions about how difficult that would be. It would have to be a group effort. Ryan, I need your financial expertise. Gabby, we'll need software development. And Ben, you've proven that we have a viable idea."

"We've supported you all along," Ben said. "Of course, we're with you."

"I know, and I appreciate it. Ryan, tell me if I'm wrong, but I should have a legal commitment. We should all have a piece of any company."

Ryan considered protesting, but Camila was probably right. Trying to start a new company with unproven technology would be hard, but beginning with a core group would make it a little easier.

"I don't need a share," Gabby said. "I'll have a piece through Ben's share."

Ryan stared at her. "That would only work if you were married. You aren't, are you?"

"No, but we will be." She turned to Ben. "Right?"

Ben had trouble speaking, but he managed to nod.

T HEY HUNG OUT AT the restaurant for a little while after eating, but Ben was exhausted, so they all got back in the van. The vehicle dropped Ryan and Mike off at Ryan's home in Albany and went on to Ben's apartment.

When the van arrived there, Gabby helped Ben out and then turned to Camila and Dani. "I'm going to stay with Ben tonight. He's going to need help."

"I'll be fine. I told you that the station has a rotating section for exercise, so living in orbit isn't as debilitating as it used to be." But Ben leaned on the van while he spoke. "Maybe a little help to get to the door, but at least there are no stairs."

"Nobody has been in there for six months," Gabby answered. "Who knows what it's like, and you're in no shape to deal with it."

Ben started to protest, but Gabby pulled him into a long kiss. When she released him, he didn't protest as Gabby helped him to his door. As Ben fumbled for his key, Camila saw Gabby kiss him again, and then the door opened, and they vanished inside.

The scene lingered in Camila's mind. No one had ever kissed her like that, although Greg had tried. She had a vision of being kissed that ardently before shaking her head in shock. In her imagination, it wasn't Ben she was kissing.

J UNE CAME WITH HOT, windy weather, but Camila hardly noticed, her attention on the approaching finals. When she wasn't in class, she was in her apartment, studying. During the warm spring, she had tried finding a quiet corner of the campus to study but hadn't done it often. Wearing a mask for an extended period was uncomfortable then; now, with the hotter weather, it was impossible.

Her last class that day was over. She closed the notes on her reader, stashed it in her purse, and joined the other students in streaming out into the corridor. Professor McGavock was standing just outside. *Waiting for me?*

"Good afternoon, Miss Lopez," McGavock said.

Camila slowed but didn't stop. "Good afternoon, Professor."

He fell in next to her as she walked down the corridor. "How are things going for you? Looking forward to summer?"

Camila tried to keep her eyes forward but couldn't stop herself from casting a side-wise glance toward the teacher. "I'm fine. Finals are keeping me busy, but I'll do OK."

"I'm sure you will." He paused for a second. "I'm aware of your situation. As I understand it, you were planning to use a new facility, but it was canceled."

"Not exactly." She looked at McGavock more carefully, trying to understand if the conversation had a point. "Professor Moreno doesn't think there will be a problem allocating space for my research."

McGavock nodded. "After Ben's success, I'm sure that's true. Of course, my offer stands."

"Your offer?"

McGavock raised his eyebrows. "To help you, Miss Lopez. I've offered my assistance several times, and the offer is still open. I think the work you will be doing is important, and I am certainly willing to help make it a success."

Now she understood. *He still wants to worm his way in. More now.* But Ryan had warned her not to offend the professor. "I remember your kind offer." She smiled at him. "Right now, I'm still concentrating on finishing the class work. It will be another year before I begin the research."

"Of course."

Camila hoped Professor McGavock would go his own way then, but he continued walking with her, slightly bent forward and holding his hands behind his back.

The stream of students had thinned, and they were almost alone. McGavock suddenly straightened. "And how are things with Professor Moreno?"

"She's been very helpful. I look forward to working closely with her when I start my research."

McGavock shook his head. "Not too closely, I hope."

Camila almost stopped. "What do you mean?"

"Perhaps I shouldn't mention this, but there have been rumors about, shall we say, her social life. I'm afraid she might try to take advantage of you."

Camila met with her advisor frequently, but Moreno had done nothing to make her uncomfortable. Professor Moreno liked to touch other people in conversation. A little massage of a shoulder, a hand on a knee. None of it meant anything, and it didn't bother Camila. Actually, she liked it.

It was an effort not to respond to McGavock in anger, but she managed to smile. "I hadn't heard those rumors, but I'll be careful. Thank you for the warning." She remembered a conversation with Dani and locked eyes with McGavock. "I should warn you in turn, though, that someone has suggested I be careful about working with you."

Professor McGavock's lips compressed into a tight line. "Well, that's absurd," he stuttered. "You certainly have nothing to fear from me. I just want to help."

The conversation with Professor McGavock had become uncomfortable. "It was nice to talk to you again, Professor, but I have to get home to study." She strode away, not looking back.

SUMMER IN TEXAS WAS boring. No one ventured outside if they could help it when the daytime temperatures rarely dipped below one hundred degrees Fahrenheit. A week after getting home, the mail brought one bright spot for Camila, a letter from Gabby.

"No one sends letters anymore," Camila said when her mother handed it to her. She slit open the envelope and took out the card inside, then gasped. "It's a 'Save the Date' card. They set a date for the wedding just before he went back to Collins."

Nina Lopez beamed at her daughter. "How wonderful. When is the wedding?"

"June fifteenth. Right after Gabby and I finish our classes."

"Well, congratulate her for me."

Nina looked at her, and Camila recognized her expression. When would her daughter find someone? It wouldn't do to tell her mother that she wasn't interested in finding someone or getting married.

NEW YORK WAS HOT and humid in August, but Texas was worse. Camila loved her family, but questions about her social life and her business plans got on her nerves after the first couple of weeks, and she was glad to get back to school. Dani and Gabby were waiting for her at the apartment she shared with Dani.

"Gabby has been staying here while you were in Texas," Dani said after they had caught up.

Gabby nodded. "But three's a crowd. I have the key to Ben's place, and he won't mind if I move there until school starts." Ben was back in space working on his research on what would be his last tour on the space station.

"How's the wedding planning going?" Camila asked.

Gabby frowned. "With Ben still on Collins, I could use more help."

"What do you need?" asked Camila.

"You're already maids of honor, and I appreciate that, but I still have to plan the wedding and pass my classes. My brother asks for help with his company, too. I know you guys also have classes, but I'm hoping you'll have some time to help me."

"We'll be glad to help as much as we can." Camila put a hand over Gabby's hand. "What about your mother? Or this mysterious brother we've never met?"

Gabby didn't answer, but her scowl discouraged any more questions about her parents.

CAMILA'S FINAL YEAR OF classes was a blur, and suddenly it was spring. Her research hadn't started yet, but Camila met with Professor Moreno to discuss her plans. Ben was back on Earth, recovering from his last stay on the space station, and her final exams were only three months away, so they had matters to talk about.

"I've got good news," Moreno said as Camila took a seat across from her. "After you submitted your plans for what you would need, I got a commitment for a space to work in for your research. It's in the basement of the old physics building, and it will probably be tight, but I think it will be adequate."

Camila smiled. "That's great. I know we thought something would happen, but it was still a load on my mind."

"The graduate student using it should be done by summer. We can set it up for you in June."

I T WAS A LONG, hot, June afternoon, but finally Gabby had her diploma. Since her thesis on her software algorithm didn't require additional research, she was done, unlike Camila, who was only finished with classes. The school held the ceremony at the field house where Dani had played hockey.

Camila had told her parents she wouldn't be coming home for the summer, and they came from Texas to help everyone celebrate, but Gabby's parents hadn't made it from their home in Pennsylvania. Even Ryan seemed to sense that the subject was best not addressed. At least her brother, who lived in nearby Schenectady, came to celebrate with her.

Afterward, they gathered at the student union to get out of the heat. Tomás and Nina Lopez beamed their joy, and Camila's brother Diego stood a little away, trying not to be noticed. Camila silently thanked God for her supportive parents and made sure she included Gabby in their conversations.

"So, let me take you all out to dinner," Tomás declared. He looked at Camila. "Any good Tex-Mex places around here?"

"I know a good place across the river," Ryan said. "I assume you mean something more than tacos and burritos."

"Absolutely. Ryan, since you know the place, can you make the reservations for me?"

Ryan nodded. "About eight?"

"That sounds good," Camila answered. "We've got to get back to our apartments and change clothes."

"And a shower," Dani added.

"Where's your brother?" Ryan asked Gabby. "He's invited too."

"He's gone already. Something about dealing with a crisis at his company."

"Too bad," Dani said. "He's a hunk."

Mike took her arm. "You know I'm right here."

Dani reached up with her other arm and touched his cheek. "Sure, babe. Just checking my options while I wait."

Mike frowned and turned red, but Gabby rescued him. "You can all meet Jürgen at the wedding. I'll make sure he sticks around for that."

Camila and Dani went back to their apartment with their parents, and Gabby went back with Ben to change. Ryan made the reservations and messaged the location to everyone.

At the restaurant that night, Tomás and Nina soon turned the conversation to Camila's plans. "You said you weren't coming back home for the summer," Tomás said. "You don't have classes."

"No, I'm done with classes. But I can start my research after the wedding. They've found a space for me."

"What about this company you mentioned?" Nina asked. "That sounds like a terrible risk."

"I have work to do before we can move forward on that." Camila glanced at Ryan, and he took the hint.

"Right now, creating a company to exploit Camila's research is just speculation," Ryan said. "We would need to get financial support beyond what I could manage, and we can't do that until Camila has evidence that her idea will work."

"But if you do start a company, you'll all share it," Tomás said.

Ryan nodded. "We've signed a legal agreement that Camila would own thirty-one percent of any company. Ben and Gabby will be married and control fifteen percent. I will have another ten percent. That leaves forty-four percent that we can offer in return for investment. Of course, if we can negotiate a smaller share for the money we need, we'll divide up the rest."

Tomás nodded. "Giving this group a controlling interest."

"Right. Camila isn't risking any money. The alternative is to try to sell her work to someone else."

"I don't want to do that," Camila said. "I want to use the Stenhouse technology for more than energy production and to do that, I need control of it."

Tomás grinned. "Some things haven't changed. You want to build spaceships."

S ENATOR LEO CORREIA REVIEWED the daily Senate news report with increasing unease. The economy teetered on the edge of a recession as the effects of the Newman-Fredrickson bill lingered. His attempt to amend the bill had failed, and failure was never good, especially in an election year. Invitations to talk shows had become infrequent, and his opponent showed little inclination to debate him while the public blamed Correia's party for the economy.

An NPR report announced a new study on the country's population. The average number of children born to couples had been dropping for a while, and according to the study, the latest drop was the largest in years. The population, especially the working people that supported the government, was dwindling, and that also weighed on the economy.

According to the report, automation advances allowed machines to fill many of the less-desirable jobs, but failed to mention the slump in the technology industry resulting from the country's obsession with privacy. Software engineers in particular were critically needed, but enrollment for university degrees in software fields was dropping, and many working engineers had left their positions.

For a politician campaigning for reelection, such situations could be opportunities. One only had to hammer home that there was a problem and assign blame to someone else. Having a solution was helpful, but, as history frequently showed, not necessary.

He needed to talk to Val. She would have suggestions for exploiting the economic effects of the drop in births.

C ORREIA TOOK SENATOR VALARIE Peterson aside when he saw her on the Senate floor. "I was hoping to get some feedback from you about my campaign strategy."

Peterson glanced toward the podium where the President pro tempore presided, but the current holder of the office was still on the floor, chatting with associates. "Sure. What did you want to talk about?"

"I need an issue for my campaign. The National Committee is frozen into the usual complaints about excessive spending and overkill on privacy concerns. I want something that will set me apart."

"Are you just fishing, or did you have something in mind?"

"There's a report today about the dropping birthrate. Critical positions are going unfilled because the people who should fill them don't exist. It's a real problem, but I need a hook."

"Your opponent will suggest loosening immigration restrictions. That could be a quick fix. Even if we could get people to have more children, it would only help twenty years down the road."

"Immigration could supply bodies more quickly, but the immigrants wouldn't have the skills we really need. It's another long-term answer at best."

Peterson shook her head. "That's not true. A lot of immigrants have skills. The percentage would be higher if we filtered applications by our needs."

"Maybe you're right, but that's my opponent's argument. I could support it as well, but that would anger the party, and wouldn't be as effective as a story of my own. He'll suggest training programs, too. I need something that's mine and isn't counter to our party's platform."

"You can't advocate for more qualified workers and asking the country to have more babies is too long term. That doesn't leave you with much, Leo."

"I don't need something that will work. I need something that the voters in my state think will work."

Peterson chuckled. "I would accuse you of being too cynical, but you're no more cynical than most of our colleagues." She paused. "All right, asking people to have more babies probably won't work. What about asking that more people have babies?"

Correia stared at her. "What?"

"At the current birthrate, a married woman would have about two babies each, creating replacements for herself and her husband. But not all women will have babies, so the population shrinks."

"OK."

"If we can't ask them to have more than two babies, suppose we ask more women to have babies?"

"Sure. We can encourage teenage sex. That would get me a lot of votes."

"If teenagers voted. But that's not what I meant. What one potential group, already looked down on by many in our society, doesn't typically have children?"

It took him a few seconds. "You mean gays? How practical is that?"

"You said it yourself. It doesn't have to work. It only has to sound good to the voters. It's become politically inexpedient to cater to racial bigots, so the best way to attract the neanderthal vote is by attacking homosexuality."

Correia shook his head. "And you called me cynical."

"I called all of us cynical. But it could work."

CORREIA FINALLY GOT HOME a little after 9 pm. Priscilla had already eaten, but she sat down with him after reheating some of the dinner and serving it to her husband.

"You look tired," she said as he poked at the food.

"It's been a stressful day, Pris," he responded.

"Bad news?"

Correia sighed. "Not really. We're in for a tough campaign, and I'm not looking forward to it."

"I thought you enjoyed campaigning. That's what you always say."

He nodded. "Not so much this time. It's going to be close, and my opponent has all the talking points."

"Has Harding said something?"

"Nothing unusual. But I need an issue to pin my campaign on. The economy is always a good issue, but it's hard for me to come up with a story I can use. I think I have one, though. The population of the United States has shrunk as parents have fewer children. That's harmful in the short term, but could be disastrous in the long term. Homosexuals rarely have children, exacerbating the problem."

Priscilla raised her eyebrows and shook her head. "You're going to base your campaign on attacking gay people because they don't have children?"

It did sound ridiculous. *But what else can I do?* "They're unpopular in much of the country already. People will believe it." He felt a dull pain in his stomach and put his fork down. "I'm more tired than hungry. I think I'll go to bed early."

Priscilla stared at him for a moment, and his pain got a little worse.

"I'll join you as soon as I clean up," she said. She took his plate and silverware and carried them into the kitchen.

CAMILA STOOD BETWEEN DANI and Kaelyn Cano, a friend of Gabby's from her hometown. Her thoughts flitted from one thing to another as the ceremony continued.

Gabby stood next to Ben in front of the celebrant, looking beautiful. The long, elegant gown fit her tall, slim body perfectly. Ben stood straight, frequently glancing at his bride, and looking as happy as Camila had ever seen him.

She was delighted for her friends, but wondered if anything like this was in her future. She had few close male friends, really only their group, and they were unavailable.

Gabby's parents sat in the front row with her older brother, Jürgen. Camila had met them before the ceremony, and they seemed normal. It wasn't apparent why Gabby didn't get along with her parents. Everyone would be on their best behavior, though.

Jürgen had seemed nice at first impression, but she was used to attracting interest from young men that she didn't detect with Jürgen. He was a handsome, athletic-looking man, perhaps five years older than her, but she felt no attraction to him either.

Camila's thoughts drifted back to her thesis research. After years of study, she could pursue her interest in the Stenhouse Field in earnest, and that was exciting. But a fear of disappointing the friends who had given her so much support was stressful.

Camila's parents, still in town after her graduation, had also been invited, and she looked for them among the attendees. She found them a couple of rows back, sitting with Ryan and Ben's parents. Her mother looked her way, probably also wondering when it would be her daughter at the altar.

AT THE RECEPTION, RYAN sat with Camila's family. Camila, Dani, and Mike were in the wedding party and sat with Ben and Gabby. Tomás introduced him to

Camila's brother, Diego, but Diego was spending more time at the buffet than at their table. Tomás and Nina Lopez were having a good time, and Ryan enjoyed their company.

Ben had introduced him to Hugo and Emma Varela briefly before the wedding. The Varelas, sitting at the next table, didn't look like they were enjoying the reception. Their son, Gabby's brother Jürgen, had wandered away, and Hugo and Emma were talking to each other and ignoring everyone else. But he could still hear them.

"He doesn't even have a job," Emma said to her husband. "His head is full of far-fetched ideas from his friends."

Ryan turned and saw they were looking toward the newlyweds. He glanced at Tomás sitting next to him; Camila's father had heard Emma Varela too.

"She can't support him, playing with computers for a living with her brother," Hugo answered. "And she'll be starting a family."

Ryan frowned and leaned toward Hugo. "He won't have a problem finding a job. Neither will she. And they'll have a share in the company we're planning and can come to work there once it's started."

"That's not a real business," Hugo said. "What's its product?"

"For now, it's strictly a research company. We hope to have products as our research advances."

"Emphasis on 'hope' to have," Emma said.

The sarcastic tone grated on Ryan, but he tried to keep his voice even. "It's a risk. Not really for them, though. If they succeed, they'll be set. If they don't, they'll still have experience that will look good on a resume."

"Then you admit this is all a wild dream of that Lopez girl, and that you're taking advantage of our daughter's talent." Emma smiled in triumph.

"For the record, they came to me." Ryan allowed a little hardness in his voice. "That 'Lopez girl' has ideas that could help the entire world. Ben and Gabby believe in her and joined her with no urging on my part."

"And you hang around with these college kids hoping to find one with an idea you can exploit," Hugo said.

Ryan could almost feel his blood pressure rising. "Sure, that's why I was there last week when your daughter graduated." He hesitated and glared at Hugo. "Where were you, by the way?"

Hugo's face reddened. "I was busy. I have an actual job."

Ryan looked at Emma. "And you?"

Emma looked down at the table. "I wasn't going to go without my husband."

"Ben has responsibilities now," Hugo said. "He needs to provide for his family."

"If you weren't so busy, you might have noticed that your daughter is very talented and quite capable of providing for herself if necessary," Ryan said.

Hugo scowled at Ryan. "I'm a good father. I take care of my children, but that means I can't be there all the time. And she already told her brother she would help him rather than get a real job."

The conversation was getting out of control. Ryan had hit a nerve with Hugo, and his repeated reference to 'a real job' irritated Ryan. But Hugo's commitment to his job reminded him of himself when Mike was young, and Ryan still had a wife. His anger faded, replaced by pity for a man who still had to learn what Ryan had discovered only in the last few years.

"She'll be fine." He smiled and turned back to the people at his table. Tomás was frowning, and Nina and Ben's parents were pretending not to have heard the exchange. He looked at Tomás with a shrug, and Camila's father nodded and turned toward his wife.

"YOUR PARENTS DON'T SEEM to be having a good time," Ben told Gabby. "A minute ago, I thought your father and Ryan were going to get into a fight."

Gabby glanced over at her parents' table. "They probably started on their usual 'she's throwing her life away' diatribe. I'm guessing Ryan made the mistake of defending us."

"Your parents don't approve of me? Or involvement with Camila?"

"They didn't even approve of my choice of software engineering. They weren't going to pay for something they thought was beneath our family. I could only afford my education because of scholarships. They feel the same way about my brother and my working with him until Camila is done."

"The world runs on software. How is that beneath you?"

"My father is a well-paid manager for a factory owned by a major manufacturer. He likes to call that 'real work.' As opposed to what I want to do, of course. To him, we're all directionless geeks. Even my brother."

"Geeks? Even his words are out of date."

Gabby chuckled, and when he turned toward her, she grabbed him and kissed him. "Forget about them, Mr. Pinto."

"Forget about who, Mrs. Pinto?"

23

A FTER FINISHING HIS BUSINESS call, Jürgen Varela decided to check out the appetizers at the buffet table. He fell into line behind someone who already had a plate three-quarters full of assorted items. As Jürgen watched, he added two empanadas to the pile and dropped the tongs he had used. Jürgen took them, causing the young man to glance at him, turn away, and turn back with wide eyes.

"You're Jürgen Varela," he said, almost but not quite stammering.

Jürgen had to suppress a laugh at the double take. "I am."

"What are you doing here?"

Jürgen waved toward the bridal party table. "My sister is the bride."

The young man looked in that direction, his plate tipping precariously. His eyes were even wider when he turned back to Jürgen. "I love playing *A World of Your Own*. I can't believe I'm meeting you."

"You know my name. What's yours?"

"Diego Lopez. My sister is your sister's maid of honor."

"Camila? I met her earlier. My sister says she's very smart."

"Uh, huh. Not as smart as you, though."

Jürgen shook his head. "I suspect both our sisters are smarter than me. I started earlier and have been lucky. They're going to do some great things."

C AMILA SAW BEN AND Gabby get up and stroll between the tables, stopping to greet the attendees. *I should be social, too.* She walked over to the table where her parents sat with Ryan.

"It was a lovely wedding," Nina told her daughter. "Wasn't it, Tomás?"

"Yes, very nice," Tomás said. "And you look even more beautiful than your friend, Camila."

Camila blushed and glanced at Ryan.

He smiled. "It's hard to judge two such gorgeous young women."

Feeling a little embarrassed, Camila turned toward the table where Gabby's parents were watching. Both were frowning, but their faces relaxed as she faced them. Did Ryan's remark offend them? "Mr. and Mrs. Varela, Gabby has told me so much about you."

Emma responded politely, and Hugo nodded in acknowledgment, but Camila felt uncomfortable. Ryan might have noticed and decided to rescue her.

"Where's Diego?" he asked.

"Probably haunting the buffet table," Nina said. "My son can make food disappear faster than anyone I know."

Camila looked in that direction. "No, not right now." She paused and glanced around the room. "There he is. He's sitting at a corner table with Gabby's brother. With a plate full of food, of course."

She saw the Varelas in the corner of her eye, and they frowned again. She hadn't entirely believed Gabby's stories about her parents, but she was beginning to.

"I think I'll go see what they're up to," she said.

J ÜRGEN ENJOYED TALKING TO the people who used his software, especially outspoken gamers like Diego. "The different scenarios are great," Diego told him. "Other games have better graphics, though."

"The program generates scenes thirty times a second, and that takes processing power. It adjusts resolution according to the power of the computer and the bandwidth of your network supplier." Jürgen shrugged and smiled. "You may need to improve one or both to get the best performance."

Diego nodded. "My computer is a few years old. And other games play slower on it than they do on my friends' computers."

"My sister is helping me improve some of my algorithms. That may help you with future upgrades."

Diego looked past him. Jürgen turned to see what had gotten his attention and saw Camila approach them.

She smiled. "Is my brother monopolizing you?"

Jürgen returned the smile and shook his head. "We were discussing a common interest. Join us."

"Gabby told me you've written a computer game. Knowing Diego, I assume that's the common interest."

"He's famous!" Diego said. "Rich, too."

Jürgen chuckled. "Not exactly. I'm the first one to use artificial intelligence to generate the entire game scenario on the fly, and that's been profitable, but I'll have imitators soon."

Camila sat down next to Jürgen. "Then what?"

"I've been investing everything I've made in growing my market. A multiplayer game will be out soon, but for the longer term, I think the key is to sell a stadium version that will take advantage of augmented virtual reality. Royalties from that would provide ongoing income."

"It sounds fascinating. And Gabby will work for you?"

"I suspect you're being polite. You don't strike me as a game player, Camila." Jürgen leaned back in his chair. "For now. I think I'll lose Gabby when you're ready to start your company. I can't say I blame her. Your work sounds more interesting."

"Now you're being polite."

"Not at all. My games wouldn't work without electricity, a lot of it, and I want a clean environment as much as anyone. If you're right, you could make an enormous difference."

He wasn't sure he was convincing Camila of his interest, although he was surprised at how much he wanted to. There was something about her that made him want to know her, even though it couldn't be more than friendship. He glanced toward Diego, who was listening closely.

"Diego, you should visit me sometime. I could show you what I'm planning for the future."

Diego's face brightened. "That sounds great. Where do you live?"

"Not far. I have a house and an office in Schenectady." He raised his eyebrows and turned to Camila. "I've got an idea. Why don't you both come visit me? Camila, you probably need a break before you start your research."

Camila looked doubtful, but Jürgen wanted to convince her. "I'll try to make it interesting, even for you. You'd be surprised how much physics is involved in generating the scenarios."

I'm pushing too hard. She's going to think I'm interested in her romantically. But he usually knew when a woman was attracted to him, and he wasn't getting that feeling with Camila. He hoped she had the same insight about him.

"Maybe," Camila said. "You're right. I should get away from my project occasionally. I can't start work until my lab space is available anyway."

"Then we should do it right away, before you get too busy and Diego goes back to Texas," Jürgen said. "How about Monday?"

Diego's face fell. "We're going home Monday morning."

"Talk to Mom and Dad," Camila suggested. "Maybe you can change your reservation and stay for another day or two."

Diego nodded. "I probably should tell them. I'll be right back." He jumped up and was gone, weaving between tables as quickly as the crowd allowed.

Jürgen watched his progress. "I guess he's as big a fan as he said."

"I think I've seen him playing your game when I'm home." Camila smiled. "I haven't played it myself."

"I don't imagine you have the time. Maybe I can get you interested, though."

"Diego always seems to be fighting monsters or other warriors. That's not something that interests me."

Jürgen grinned. "But that's what interests Diego. For you, I would suggest something much different."

24

C AMILA BOOKED AN AUTOMATED taxi to take them to Schenectady on Monday morning. It wasn't a long ride, and Camila had ridden the route before, but she saw the passing landscape differently. The ailing trees and rundown buildings had been the norm before, but now she saw it all as Ben probably saw it. Global warming and the cost of fighting it had stressed the capabilities of government. Too many politicians had fought the efforts, resulting in a tension that only made attempts to address the problems less effective.

She forgot that as the car pulled into the parking lot in front of Jürgen's company, a suite in a medium-sized office building. Diego jumped out of the car and charged to the door, but then hesitated and waited for his sister to catch up before they both entered. Inside, a friendly receptionist greeted them and notified Jürgen that they had arrived. He came out less than a minute later.

"Come on back," Jürgen urged after they exchanged greetings. "I can't wait to show you everything."

He's as enthusiastic about creating his games as my brother is about playing them. Diego followed Jürgen eagerly, and Camila trailed behind them.

Jürgen led them beyond the small reception area to a larger, dimly lit room with four workstations. Two were in use; a man and a woman worked on large screens filled with text and symbols. "Chuck and Gianna are working on the multiplayer version that we hope to release in a couple of months," he told them.

The two programmers looked up briefly, smiled, and turned back to their workstations.

"Mostly we're debugging now while our marketing firm prepares the launch. A stadium version is in late design. Gabby was working on that until other priorities interfered." Jürgen moved toward a door on the opposite wall. "When we need a break, we go into the next room."

"What's in the next room?" Diego asked.

Jürgen grinned. "That's where we test our software." He opened the door and beckoned Diego to follow.

Camila followed them into an unoccupied room with two large computer screens, both displaying the logo for V-World, Jürgen's company. Camila had seen the same display on Diego's computer, but the graphics quality was noticeably better here.

"Have a seat and log in," Jürgen told Diego.

"It will know me?" When Jürgen nodded, Diego sat down at one of the computers. "Log in Diego Lopez."

A light on the screen flashed. "Recognizing Diego Lopez of Pearland, Texas. Welcome to V-World, Diego. Do you want to go to Hazard Planet or create a new setting?"

"Hazard Planet."

"Hazard Planet is loaded. Do you want to resume your existing scenario or start a new game?"

"Existing."

"Resuming scenario." A dense forest of bizarre trees appeared on the screen.

Diego stared at the display for several seconds. "This is fantastic. It looks so much more real than on my computer. It loads faster, too."

"The game adjusts to the capabilities of your machine," Jürgen said. "These are state-of-the-art computers with expensive graphics systems and high-speed direct network connections."

Some of the trees moved, accompanied by crashing noises. Almost instantly, Diego's attention was completely on the screen.

Jürgen stepped back and smiled at Camila. "That should keep him busy for a while. There's another computer. Why don't you try it?"

Camila looked at Diego's screen, where an enormous creature lumbered out of the forest. She could only describe it as looking vaguely like the child of a bear and a porcupine. An arm pointing a complex weapon extended out from one side, presumably controlled by Diego. "This isn't my idea of entertainment."

"Not a problem. Sit and I'll show you."

Jürgen seemed so sure that he could impress her, and Camila swallowed her doubts and sat down at the second screen. "Log in Camila Lopez."

"Camila Lopez is not a registered user. You must open an account with V-World," the computer said.

Jürgen bent toward the screen. "I authorize it."

"Thank you, Jürgen. Camila Lopez, you are now a registered user but do not have any existing scenarios or settings. Do you want to create a setting?"

Camila looked up at Jürgen, unsure of how to proceed.

"Create a generic setting based on Schenectady, New York," Jürgen said. "We'll add details as we go."

"Setting confirmed and level set to beginner. Please describe a scenario."

"Camila would like to form her own company to manufacture a power-generating device using innovative technology she developed in getting her doctoral degree in physics. She will need a facility and employees."

"How many employees?"

Camila shrugged. "One hundred."

"Employees will be classified as I add them. Besides space for them, how much space will you need for operations?"

Camila saw why Jürgen had thought he could interest her as much as Diego. If the simulation she was developing was as complex as Diego's games, the software could even be useful.

With Jürgen's help, she spent the next hour building a factory, lining up investors, adding details, and specifying activities. The company grew and began marketing a product. The program tracked expenditures and income, displayed in detail when she "worked" in her spacious office. When she "attended" meetings, her computer-generated characters interacted with her. All was displayed in graphics indistinguishable from real life.

"You developed all this with just a few people?" Camila asked.

Jürgen shook his head. "We licensed graphics and AI software from other companies and added custom code to link them. Royalties on those products take up a substantial portion of our revenue, but it's worth it because the licenses include exclusive rights for software in games such as mine. That will at least slow down any competition." He paused and looked at Camila's screen. "My company adds the interface that uses the AI software to generate the images."

Camila nodded and turned her attention back to the game. After the initial costs of startup, her fictional corporation approached the break-even point after five simulated years of operation. The stock price rose, and the computer reported increasing interest from government and media.

Then, over two weeks, sales of her product dropped, and the stock price plummeted. "What happened?" Camila asked Jürgen.

"Computer, why have sales dropped?" Jürgen asked.

"Several existing large companies have developed products using the same technology and can market their products more successfully. Analysis suggests that Camila should have patented her product and that she will soon be out of business."

Camila stared at the screen, then at Jürgen. She shook her head and struggled to find something to say.

Jürgen chuckled. "Your mistake was more basic than that. You didn't hire any legal help or hire a CEO to run the business for you. Your inexperience with business sank you."

Diego laughed, and Camila faked a dirty look at him. "Is this game always so hard?"

Jürgen's smile faded, and he looked sympathetic. "You were playing at the beginner level, the default for first time users."

Camila looked back at the screen. "Thank you for a very educational experience."

"You're welcome, Camila," the computer replied. "Please come back again."

Jürgen's smile was back. "There's a small restaurant in the building where we usually have lunch. I think my staff should be getting hungry by now."

THE RESTAURANT WAS REALLY just a small sandwich shop with a few tables, but the woman taking their order greeted Jürgen by name. Rather than crowd around one table intended for four people, Diego sat with Jürgen's staff at one table while Camila and Jürgen took another.

The two programmers, Chuck and Gianna, were older, but the receptionist, Paige Murphy, was about Diego's age. Her receptionist duties were temporary; she explained that Jürgen had hired her as a software tester, but she was happy to fill in up front until the updated version was ready for test. Although Chuck and Gianna were friendly and added to the conversation while they ate, Diego focused on Paige.

"It sounds like you have a great job," Diego told her. "How did you get it?"

Paige shrugged. "Lucky, mostly. I finished an associate degree in computer programming last year and saw Jürgen's ad. Chuck and Gianna were already helping him finish up the first version of *A World of Your Own,* and they needed somebody to test.

They also needed a receptionist, and neither was going to be a full-time job. I think Jürgen hired me because I would do both."

"Wow."

"So, what about you? Are you in college?"

"I just finished a three-year program. When I get back home, I'll be looking for a job as an electrician." Diego paused. "Maybe I should have gone into programming."

"You're a gamer, though. Right?"

Diego smiled. He was proud of the scores he had achieved with various games. *A World of Your Own* was just the most recent. "Since I was six."

Paige grinned and turned toward the other table. "Hey, boss. Could we use an electrician who is also an expert game player?"

Jürgen raised his eyebrows and looked in their direction. "Diego? Maybe. We should have at least one more person to test the multiplayer version when it's ready."

"But he lives in Texas," Camila protested. "If he stayed here, he would have to find a place to live."

"He could test the software remotely," Chuck suggested. "Having a remote tester would be a good thing, actually."

Diego looked at his sister, but she just shrugged. He thought about asking how much the position would pay, but he didn't really care. He would be happy to do it for free.

25

C AMILA HAD PAUSED INCOMING calls while they were at lunch. When she checked her phone after they returned to Jürgen's office, she found a message from Professor Moreno.

"My lab is ready," she told them. "I should get back and check it out."

Diego frowned. "I suppose I should go back with you."

Camila guessed he wanted to go back to his game. "You're going back to Texas tomorrow. You have packing to do."

Diego didn't look convinced, but he nodded. "All right."

"I can understand you wanting to rush back," Jürgen said. "But maybe you would be free tonight?" He looked her in the eyes, not moving. "Dinner at a good restaurant and theater, or something more to your taste, afterward?"

The invitation wasn't unexpected, but it brought back doubts about Jürgen's intentions. It would be impolite to refuse, and she did enjoy his company. "That sounds wonderful."

C AMILA'S LABORATORY WAS IN an older building near Professor Moreno's office in the main physics facility, but the message from the teacher had suggested she go to the office first. Moreno rose from behind her desk as Camila entered the room.

"How was the wedding?" Moreno smiled, took Camila's hand in both of hers, and squeezed it.

"It was wonderful, but I don't think I ever stopped thinking about my laboratory, knowing that it would be available soon." *Although I didn't think about it when I was playing Jürgen's game.*

"Of course, it's just a room right now. You have a desk and a computer workstation, but not much else. The equipment you requested should be delivered tomorrow. Fortunately, we already had most of it, left over from Ben's project."

Camila nodded. "Eventually, I have to build a proof-of-concept generator, but I'll be working on its design for a while. I'll need a place to do the tests, but that's months away."

Moreno walked toward the door. "Right. Well, let's go take a look."

They walked the short distance to the building housing her laboratory. The room was in the basement, several doors down from the stairway, a windowless room roughly the size of a moderately large bedroom. As Moreno had promised, there was a desk and two chairs, a two-drawer filing cabinet, and a small shelf unit. A small computer processing unit sat on one corner of the desk, connected wirelessly to a large monitor and a keyboard that supported spoken and typed commands. A workbench with a row of power sockets and more drawers lined one wall. Everything was clean except for a few stains on the tile floor, probably remnants of the previous occupant's activities. Camila hadn't expected anything more, but it still felt anticlimactic.

Moreno held out a key ring with two identical keys. "You'll probably want to lock the door after you've moved your things in."

Camila took the keys. "I'll start doing that tomorrow."

C AMILA WAS ENJOYING HERSELF. Jürgen was pleasant company. He asked intelligent questions and seemed to be genuinely interested in her work. He was friendly and attentive, and she was sure he was enjoying her company as much as she enjoyed his, but for the first hour, he showed no sign of anything more than friendliness, and Camila wasn't used to that. It didn't bother her, but she was curious.

Dinner was over, and they only waited for the check. The meal had been wonderful, and she sat back and relaxed. Neither spoke for a minute, but then Jürgen turned to her.

"Do you have a boyfriend in Texas?"

Here it comes. She felt disappointed, her hopes of just having a pleasant time with a colleague apparently dashed. "No, I have no time for them." *That came out harsher than I intended.* She looked at Jürgen to see if he was offended.

But he only chuckled. "Just making conversation, Camila. That wasn't the beginning of a pass."

Unsure what to say, she only stared at him with eyebrows raised.

"You're a beautiful woman, Camila, but I have no romantic interest in you." Jürgen's smile was gentle and expectant.

"I don't understand."

"Camila, I think we could become great friends, and I hope we do. But I'm gay."

"Oh." Camila's first reaction was shock, but it quickly gave way to relief. She blinked and looked at Jürgen. "I would like to be friends, too."

JÜRGEN HAD GOTTEN TICKETS to a show in Albany, and they stopped for a late dessert afterward. It was well after midnight when Camila got back to the apartment she shared with Dani.

Jürgen hugged Camila at the door. "I hope you had as good a time as I did."

Camila smiled. "I did." She opened the door and entered the apartment as Jürgen walked back to the car. Dani was in the living room.

"So, did you have a good time?" Dani asked.

"Jürgen took me to his favorite restaurant and a show downtown. I had a wonderful time. He's a great host."

Dani grinned. "It sounds like there's a spark of something there."

Camila wanted to tell Dani everything, but she hesitated. "I think we can be friends."

Dani's face scrunched into a puzzled frown. "Just friends? You had a wonderful time, and the guy is a hunk. What aren't you telling me?"

"I really like him and enjoy his company, but it's not a romantic attraction."

Dani stared at her for several seconds, but then she nodded. "And Jürgen feels the same way about you."

"Yes, I think so."

AS PROFESSOR MORENO HAD promised, the instruments Camila would need were delivered the next day. Camila spent an hour inspecting everything, flipping power on and doing any checks that she could perform quickly. It wasn't necessary, and she knew she was being compulsive about it. Then she unpacked a box of books not available on her reader and arranged them on the shelf.

One critical piece of equipment was missing, but that she would have to build herself: the Stenhouse Field generator that was at the core of her thesis project. The equipment she had just installed was either normal for a physics lab or needed to help make the generator.

After an hour, she had tinkered with everything she could. She sat down at her workstation and opened her reader to transfer her work to the new computer.

Her phone announced a call from Jürgen. "Let's have lunch," Jürgen said.

Camila hesitated. Jürgen said he was gay, but he acted as if he were interested in her. His approach was better than Greg's had been, but was it only friendship? *I guess it won't hurt to find out.* "Sure," she said.

Jürgen must have noticed her reluctance. "I think we're friends, and that's all we're ever going to be. You know that, right?"

Camila nodded. "I thought so."

"So there shouldn't be anything unusual about friends having lunch together." Now Jürgen hesitated. "There's something else, Camila."

Relief put a smile back on Camila's face. "What?"

"I need to talk to you in private. I don't trust the phone."

"OK. Where?"

"Can we meet at your lab? I'll bring food with me."

Sounds mysterious. Camila didn't know how to respond and tried to make a joke. "It sounds like I'm a cheap date. Are you having financial problems suddenly?"

His reply didn't help. "Not yet."

26

CAMILA GAVE JÜRGEN DIRECTIONS to the physics building and her lab in the basement. An hour later, he arrived, bearing a paper bag carrying containers of Mexican food. Camila moved her workstation aside, and they laid everything out on her desk.

"So why the secrecy?" Camila asked as they sat down to eat.

Jürgen glanced at the door with a worried frown. "Everything I said about wanting to be friends was true. You know that, right?"

Camila nodded. *Where is he going with this?* "I thought so."

"That's still true. But after this morning, it's not the only reason I want to be friends. Being seen with you socially would make me feel safer."

"Safer with me? I don't understand."

Jürgen pulled a phone from his pocket and tapped the screen. Then he handed the phone to Camila. "Play this video."

SENATOR LEO CORREIA STOOD behind a forest of microphones with fellow Senator Valerie Peterson standing behind him.

"We have a problem in this country," he began. "Our factories have cut production and critical occupations such as teachers and service employees suffer from a lack of people to fill positions. The root cause is, of course, the declining birthrate among Americans. Meanwhile, economic rivals such as China have all the human resources they need to outcompete us.

"We must address this problem. Obviously, we can't force women to have children. My colleagues across the Senate aisle will tell you that was tried a couple of decades ago and failed. What we can do, however, is create new incentives for the stable, traditional marriages that are America's greatest resource.

"Senator Peterson and I are introducing a bill to address this issue. The bill will place a tax on single adults and couples in nontraditional marriages. Further, it will put new requirements on applications for marriage licenses between two persons not identifying as a man and a woman.

"This is not an attack on gay marriage. Homosexuals have as much right to choose their life partners as heterosexuals. However, they must also bear the responsibility of being unproductive in the most important contribution we can make to our country."

JÜRGEN TOOK HIS PHONE back. "That won't result in more births," Camila said. "Being friends with me won't prevent this tax he's talking about."

"You're right. But it's not the tax that concerns me. Homosexuals are already threatened in this country. Speeches like this, blaming us for the problem, will make things worse." He smiled, but it was weak. "Being seen with a beautiful woman will mark me as a straight man and not part of the problem."

Camila shook her head, but Jürgen was probably right. It had taken hundreds of years for Americans to get mostly over their xenophobia about people with different skin color, and a lot of that progress had taken place in the last twenty years. People who hated gays were still all too common.

She needed to talk about it more, but she understood enough already. "Next time, we'll go out for lunch."

BEN AND GABBY GOT back from their honeymoon two weeks later and showed up at Camila's laboratory the day after the July Fourth holiday. Camila heard them down the hall and walked out to greet them.

"So, how was the honeymoon?" she asked as they moved back to her room. "You stayed in Yellowstone the whole time?"

"Two weeks," Ben confirmed. "We didn't do much the first week while we got used to the altitude."

"Don't listen to him." Gabby took the second chair and Ben perched on an empty spot on Camila's desk. "We did plenty without leaving our room."

Ben blushed, but he continued. "We hiked all over the park once we were used to the thin air."

"It was beautiful," Gabby added. "And I had Ben to explain everything we saw."

Ben glanced around at the blank walls. "We took a lot of pictures. This place would benefit from some framed photographs. I could blow up pictures I've taken in the Adirondacks, too. With some nice frames, these walls would be friendlier."

Camila nodded. "You're right. I just haven't gotten around to decorating."

"I assume you've been working here since the day after the wedding," Ben told Camila. "Have you anything new to tell us about?"

"I've been busy, but I'm not sure how much I've really accomplished," Camila said. "I got more data on Stenhouse's work from the Advanced Physic Institute where he did his research. I need to design a Stenhouse Field generator and get it built, and I've learned something about how to do that. You could copy his design mostly, but I can't do that until I know more about how we're going to harness the field to get usable energy." She paused and shrugged. "It's easy to talk about pulling atomic nuclei apart, but actually working with that kind of precision at such small scales is not trivial."

"I've been thinking about that," Ben said.

Gabby looked at him, surprise and shock on her face. "You have? When did you have time?"

Ben blushed again, but he tried to recover. "When you were asleep, darling."

Camila grinned. "So, did you come up with anything with this thinking?"

"Maybe. I took our speculation about the edge effect one step further. When you generate the field, you're warping space. The result is a boundary where nuclei seem to fission. If you could take that idea a little farther, what would happen to atoms inside a closed Stenhouse surface?"

"I'm not sure." Camila frowned. "Would there still be a Higgs field within the area? If not, any particle within would be massless. Would the particle convert the mass energy to motion? Probably at the speed of light?"

"Productive research on that interface would be enough to satisfy your thesis requirement," Ben said. "If we could do it. Stenhouse only warped space a little in his experiments. The same with mine."

"A stream of protons and neutrons, accelerated to near the speed of light, would make a great spaceship drive," Camila said.

Ben smirked at her. "Yeah, that too."

Gabby changed the subject. "My brother says you've been seeing each other."

"We have."

Gabby frowned. "I was a little surprised at that."

Ben looked at his wife. "Why? Why wouldn't Jürgen be interested in Cam?"

"We're really only friends," Camila said. "We understand each other."

Gabby nodded, and her frown faded. "Good."

Ben shook his head. "Am I missing something?"

Gabby patted his cheek. "Don't worry about it, sweetie. It's just girl talk."

27

"**I** could use your help," Camila told Gabby. It was another Friday night, and everyone was waiting for the pizza.

Gabby nodded. "Sure. What do you need?"

"I'm working on the design for my generator. Ben copied Stenhouse's design, testing it without gravity affecting the data. I need to do more, shaping the field and controlling the breakup of heavy atoms well enough to generate reliable power. I've gotten time on the school's quantum computer, and I think the algorithm you developed could help simulate different field configurations."

Gabby glanced at Jürgen. "My brother keeps me pretty busy, but I can find time to show you how to apply my work."

"Thanks. That's one problem addressed."

"What else?" Gabby asked.

"I'll be creating a flood of protons and neutrons that will have to be controlled. Protons have charge and a reactor could manipulate them with an electric field, but neutrons have no charge. They could damage the material around the reaction chamber. Nuclear power plants have the same problem."

"I can't help you there," Gabby said.

"I bet you can find somebody in the Materials Engineering Department to help," Ben said. "Somebody needing a thesis subject might appreciate the opportunity."

"To work with Camila, he might not need a thesis to motivate him," Mike said. That earned him an elbow from Dani.

Ben was right. When Camila talked to Professor Moreno at their next meeting, she helped find Rafael Rodrigues, a student who was considering graduate school. He liked the idea of investigating materials resistant to neutron emissions, and, with the added benefit of working with Camila, accepted immediately.

IT HAD BEEN OVER two years since the Newman-Fredrickson Privacy Act became law. The bankruptcies for the losers were complete. Companies that could adjust survived. Some consolidation took place among companies that exploited personal information, and government agencies explored other ways to monitor citizens. As with prior economic upheavals, the country was recovering, and the previous disruption was almost forgotten.

But the new standard made good polling data impossible to gather. Senator Correia found it difficult to assess his chances in the quickly approaching election and began relying on conversations with his constituents at campaign rallies. The information he got was more nuanced, but less relevant statistically. His remarks about birthrates and homosexuals were appreciated by the people he talked to, but he was still worried. His wife warned him that the people at his rallies were not an accurate cross section of the people in his state, and he knew she was right.

Contact with constituents was the best information he had, but getting a wider range of opinion wouldn't hurt. "What do your voters think about what I've been talking about?" he asked Senator Peterson.

"Most of them like it," Val told him. "I don't think that helps you judge your state, though. Texas doesn't like the other side's ideas about solving the birthrate problem by making immigration easier."

Correia nodded. "My supporters don't like immigration much either. I could add more anti-immigrant statements to my speeches. Anything I can do to make New Yorkers feel the same way as Texans will help."

"It's worked to get votes in the past."

THESIS RESEARCH INFORMATION WAS available to all the college's teachers, and Professor McGavock stayed abreast on the work of many thesis projects. The college had a studio for the faculty so they could record lectures for remote students, and professors used the facility for podcasts, disseminating information about their research or commenting on other subjects relevant to their field.

In mid-August, McGavock recorded a lecture from one of his astrophysics classes. He had considered approaching Camila again to offer help in the hope of becoming involved in the project, and his work that day inspired him to use his knowledge to create a podcast about the project. Excited, he went back to his office to write up a script. The next day, he was ready to record a short talk.

"There are many interesting research projects being carried out here. One with exceptional promise is a thesis effort, titled *The Application of the Stenhouse Field to the Exploitation of Nuclear Energy,* with the potential to revolutionize the energy economy of this country and even the world.

"Camila Lopez's research builds on the work of another student of mine, Benjamin Pinto, who confirmed the existence of the Stenhouse Field for his PhD. Several other students, including Ben Pinto, assist Miss Lopez in her efforts.

"The project is based on the work of Arthur Stenhouse, the lead scientist at the Advanced Physic Institute in Spokane, Washington. Stenhouse discovered the phenomenon dubbed the Stenhouse Field and showed how it could warp space. His work provided valuable clues about a theory that would unify quantum mechanics and gravity, but did not develop any applications.

"Our university, through the work of Miss Lopez and advice from our faculty, hopes to change that. Miss Lopez believes the Stenhouse Field can split the nuclei of heavier elements and release the binding energy. If this sounds like nuclear fission, that's because that is exactly what it is. The difference between the Lopez proposal and existing nuclear fission is in the material used as fuel and, perhaps, the energy yielded.

"Existing nuclear fission requires radioactive materials such as uranium and plutonium. These elements spontaneously decay at a rate that a nuclear power plant can control, but release only a small fraction of the energy available and produce problematic waste products.

"Theoretically, the Lopez project will use any element heavier than iron to produce energy. Lead, for example, a material much cheaper and easier to handle than radioactive elements. The energy output is potentially much higher, releasing the energy in the fuel by repeated decay until the fuel has been reduced to iron. The resulting waste would not be radioactive, eliminating the problem of disposing of hazardous materials.

"Work has been going on for only a few months, and it is too early to estimate the chance of success. I will watch the project's progress and intend to report on it in future podcasts. Thank you for listening."

ARTHUR STENHOUSE STARED AT the computer screen, a sour expression revealing his opinion of the physics paper displayed there. Yet another young physicist attempted to unify quantum mechanics and general relativity and, as usual, completely ignored the implications of his work and the existence of the Stenhouse Field. The adolescent scientists that dominated physics weren't interested in the discoveries of a middle-aged colleague.

He knew he should continue his work, especially now that another researcher had confirmed his discovery, confounding those who had considered it a dead end and ignored it. Even now, though, the general feeling was that it had no applications, and the complex mathematics that obscured the clues it offered on the nature of the universe were beyond the skills of the neophytes who authored papers like the one on his computer.

He had been one of them once. Then he had dedicated decades to prove that a vague idea represented a real phenomenon. He was fifty-four years old when he generated the first Stenhouse Field. His expectations of fame and a Nobel Prize faded away in the years that followed. Now, sixty-one years old, he was an ancient, forgotten relic, too tired to continue the grueling effort that advancing would require.

"Close document," he told the computer. He hadn't looked for mentions of his name since the day before and decided to do it then. "Open Stenhouse name search."

Usually, the program would find one or two citations where his name appeared. Today, three appeared and one looked promising. Some college professor from the east coast had made a podcast and mentioned his name. Curious, he selected the link to the video and his screen displayed Professor McGavock.

"Play," he ordered. When the short podcast ended, Stenhouse played it again. A hint of a smile escaped from between his lips. A graduate student, attending the same college as the researcher who had confirmed his work, had a plan for harnessing the Stenhouse Field to release energy.

"Give me information on physics Professor McGavock." There wasn't much—the limitations of the government's interference in information access, no doubt—but it had a phone number.

He reached for his phone.

W HEN CAMILA WENT TO bed the previous night, Dani had not returned from a date with Mike. It wasn't unusual for her to return late, but she usually had breakfast with Camila the next morning. Dani was an early riser, regardless of how late she was out the night before.

When Dani didn't appear, Camila knocked on her door. Dani didn't answer, and she peeked in, noticed the bed hadn't been slept in, and closed the door again. Dani had never stayed out all night before, which worried Camila. She had left her phone in her bedroom and went back to retrieve it.

It was under her reader. She had read in bed for a few minutes before falling asleep, so she hadn't noticed it blinking frantically to announce a message from Dani. With some trepidation, she started the message and immediately smiled. On the screen, Dani sat on a couch, clearly excited.

"Cam, I won't be home tonight." Dani glanced to her left. "Mike has proposed. We're going to be married right after I finish my MBA."

Mike moved into view and grinned at the phone before kissing Dani on her neck. Dani pushed him away, not very hard, with a giggle. "I'll give you all the details tomorrow." She turned toward Mike, and the message ended.

Probably not all the details. Camila smiled and put her phone down. She expected an announcement eventually, and wasn't surprised they were waiting until Dani graduated, but she also couldn't help thinking it was about time.

After a solo breakfast, Camila left the apartment and walked to the campus, invigorated by Dani's news. For an early morning at the end of October, it was already warm, and she had to force herself to slow her pace. Perspiration still made her face itch around the edges of her mask.

The building's air conditioning hit her like a cold towel, and she ducked into a restroom to wipe her face and check her makeup. In her office, she spent her first few minutes reviewing a report Rafael Rodrigues had written on neutron-resistant materials. He was investigating current possibilities in his spare time as he studied for his bachelor's degree and worked on his own thesis proposal.

Her phone lit up, and she hoped it was Dani, wanting to schedule lunch so that they could talk, but it announced a call from Professor Moreno.

"Are you free to meet this afternoon?" Moreno asked.

"Sure. Is there a problem?"

"I don't think so. McGavock wants to meet, and I hate to put him off since he's on the thesis committee."

Camila nodded. "I can make it. Where and what time?"

"He wants to do it here. The second-floor conference room here at two?"

Camila wanted to know what McGavock wanted to talk about, but Moreno would have told her if she knew. "I'll be there."

They ended the call and Camila returned to her work, but the energy she had gotten from Dani's news had disappeared. The astrophysicist wouldn't leave her alone. She considered calling Ben or Ryan, but didn't. *I need to stand on my own.*

W HEN CAMILA ENTERED THE conference room, Professor Moreno was already there, talking with Professor McGavock. Next to him, a man she didn't know sat at the table. He looked up at her as she entered with a disgruntled expression that made Camila wary. He was probably in his sixties, heavyset, and dressed in a worn suit. She couldn't be sure of his height because of his seated position.

McGavock rose. "Camila, I have a surprise." He waved a hand at the man next to him. "This is Doctor Stenhouse. When he heard about your work, he contacted me, and I invited him to come here from Washington."

Stenhouse stood and extended his hand. "Hello, Miss Lopez." He smiled.

At least the change in expression looked like a smile. "Good afternoon, Doctor," she answered.

"I was hoping to meet Doctor Pinto as well. He couldn't be here?"

"He wasn't available," Professor Moreno answered.

"That's too bad. I wanted to thank him for confirming my discovery." Stenhouse paused. "But I was more interested in coming here when I learned somebody was researching an application for my work." Stenhouse released her hand and sat down again.

Camila also took a seat while she tried to form a reply. It was all she could do to maintain a cool demeanor sitting across from the scientist who inspired her work.

"Perhaps you could tell me a bit about your idea," Stenhouse continued.

Having Doctor Stenhouse come to talk to her was exciting and validating, but Camila was cautious. She might have felt differently, meeting Stenhouse only, but McGavock had brought him. She looked at Moreno, and her advisor's expression didn't encourage trust.

Camila had to say something, and most of her work was public anyway. "Of course. My original idea was to use the Stenhouse Field to compress matter and initiate fusion."

"I thought of that myself," Stenhouse said. "It was impractical, however. Pressure alone wouldn't be enough."

"Yes, I realized that. I then considered fission, and that's the current direction of my research."

"I didn't see a good way of configuring the field to tear apart individual atoms." Stenhouse frowned and looked disappointed.

Camila spoke quickly. "I'm researching how atoms react to the spatial stress at the boundary of the Stenhouse Field."

Stenhouse leaned forward again and nodded slowly. "What have you found out?"

"It's still too early. I'm working on designing a Stenhouse Field generator for now. I won't know if I have a workable concept until I can perform experiments."

Stenhouse moved back. "That's very interesting. Perhaps we can stay in touch so that you can update me when you have progress."

"I'm curious," Camila said. "Thousands of doctoral students work on thesis projects in relative obscurity. How did you find out about my work?"

Stenhouse glanced at McGavock. "Professor McGavock's podcast, of course. Isn't getting some attention why you had him publicize your project?"

Camila stared at McGavock with narrowed eyes. "Of course. I don't know why I didn't think of that." But she was confused. *What podcast?* She didn't remember asking for publicity from McGavock or anyone else. Then she nodded. Her former professor was again trying to get involved in her thesis work. She resolved to tell Stenhouse as little as possible.

"Do you have any advice for me, Doctor?" she asked.

"Not immediately. I hadn't considered interactions at the field boundary. That would explain the glow I observed in my work, though."

Stenhouse and McGavock both tried to draw more information from Camila, but she put them off, explaining more than once that she had only started the research a short time before, and didn't have any concrete results to report.

Stenhouse asked about Ben, expressing his pleasure at having his work confirmed, but McGavock pushed the conversation back to Camila's work. Another hour went by before they gave up pumping her for more and left.

29

CAMILA HAD HANDLED THE meeting with McGavock and Stenhouse well enough, but it left her shaken. It was Friday, and she could talk to Ryan and Ben at their usual meeting. The group would celebrate Dani and Mike's engagement announcement, too, and Camila hoped her concerns wouldn't dampen the mood.

Gabby invited Jürgen, making a group of seven people. When they assembled that night, they were recognized as regulars, and the pizza restaurant staff set them up with two tables pushed together in a corner of a room off the main space. They didn't have it to themselves, but it was a little quieter at first. When Dani told everyone about the engagement and showed her ring, she and her friends raised the noise level significantly.

Camila waited until everyone had eaten before broaching the meeting that afternoon. Everyone was relaxing over mugs of beer, but that changed as she told them about McGavock and Stenhouse.

"It's probably a good thing that you didn't tell them much," Ryan said when she finished.

"Professor McGavock made me nervous. Doctor Stenhouse could be helpful, though."

Ryan shook his head. "He could if he wanted to. Maybe I'm being paranoid, but we know McGavock would like to be a part of this. What do we know about what Stenhouse wants?"

Camila's mouth twisted and her eyebrows moved close together. "I don't know. He flew here all the way from Spokane."

"Invited by McGavock. That makes me want to be cautious in dealing with both of them." Ryan paused. "I think we should consult a patent attorney about what we can do to protect your work. Otherwise, someone like Stenhouse could steal the idea."

"If you think that's necessary," Ben said.

"I don't know. That's why I want to talk to someone who does know."

"A N AGENT FROM THE Privacy Protection Commission will visit later today," Jürgen told his staff.

"Why?" Paige Murphy asked. "They've never worried about us before."

Jürgen shrugged. "It's a relatively new agency. I don't think we have anything to be concerned about."

Jürgen didn't want to worry his employees, but he was more uneasy than he said. The PPC had been created to enforce the provisions of the Newman-Fredrickson Privacy Act. Jürgen knew it grew rapidly during the year since its establishment, but he was surprised that his gaming software had caught their attention.

"She should be here around two. Just bring her back to my office."

"Okay, boss," Paige responded.

Within a minute of two o'clock, Paige brought the woman into Jürgen's office off the room where the programmers worked. She was about Jürgen's age, and would probably be considered attractive to most males. When Jürgen shook hands with her, she held his hand a little longer than necessary.

"What can I do for you?" He waved at a chair next to his desk.

She smiled. "My agency is talking to companies that might be inadvertently acting counter to recent legislation on the right to privacy."

Jürgen nodded. "I assumed as much. I have had lawyers check out our operations, and they have assured me my program fully complies with the relevant regulations."

"As I understand it, your software builds scenarios by analysis of outside data."

"Yes. But only public domain data and very little of that. The program relies much more on simulation using known science. It collects no data from its users."

Her smile faded, but only slightly. "Don't your users have to enter information for authentication?"

"They give a username, not necessarily their actual name, when they register their copy. The software can remember a user's face that is associated with the registered user name, the only authentication necessary. Users without a camera can use a password, but we discourage that. Too many people use the same password for other accounts, which can lead to complications. As I'm sure you know, passwords have been a tremendous problem historically."

"There are more modern ways to authenticate."

"Sure, and we considered retina scanning and other biometric methods. But they would require additional hardware not available on all computers, and we didn't want to add to the buyer's cost. We might add one or more of them as an option in future versions."

"You have advertised a future multiplayer version. Won't that require additional user information?"

Jürgen leaned back and considered the question for a moment. "Players would have to exchange user names, but no more. Of course, because of the flexibility of the game, a user could reveal personal information, but it would strictly be the user's decision."

The woman smiled. "I would be tempted to reveal personal information if I were in a game with you."

Hell, she's flirting with me. Jürgen forced a tiny smile and nodded slightly, hoping she would see that as neutral.

It wasn't the response she hoped for. She frowned and her voice turned cooler. "You use artificial intelligence to create scenarios. Couldn't your software use that to violate a user's privacy?"

"I've licensed AI used by many other applications. Any privacy violations would come from the sources it uses and those are limited to noninvasive information."

"We'll need to verify the sources you use to generate your scenarios."

"I believe that information has already been filed with your agency, as required. I'll check, though, and make sure our disclosures are current."

THE PPC AGENT ENDED the interview soon after. Jürgen didn't offer to escort her to the lobby, but it was a small suite, and she didn't have any difficulty in finding her way out. She didn't expect him to show any significant interest in her, but his indifference was annoying.

She stopped at Paige's post in the lobby. "Your boss is a good-looking guy," she said when Paige looked up.

Paige grinned. "He is that."

"Are you dating him?"

Paige shook her head. "No, of course not."

But the grin disappeared, and the woman regretted the personal question. What difference did it make?

30

JÜRGEN WAS STILL THINKING about the visit the next day. Paige had told him about the conversation as the PPC agent left, which only confirmed his belief that she had been flirting with him. Most men would have found her attractive, and responded at least a little. Perhaps keeping the interview so firmly professional had been a mistake.

Senator Correia's speech had unnerved him, and the idea that an employee of a government agency might suspect he was gay increased his fears. He wasn't doing anything illegal, but that was no protection against bigots if he were outed.

Gabby came into the office. "Hey Jürgen, I'm going to leave a little early today. Ryan has called a meeting of the group tonight."

"Sure, no problem."

"I'll probably leave about five. Camila asked Professor Moreno to get permission to use a conference room in her building. I guess it's going to be all hands on deck."

Except for Gabby and Camila, Jürgen didn't know the people in the group well, but some social time could distract him from worrying about the PPC. Besides, his new brother-in-law would be there, and getting to know him better was a plus.

"Do you think they'd mind if I came, too?"

Gabby shook her head. "I'm sure they wouldn't. And Ben said there would be beer and pizza."

Jürgen grinned. "Then I'm definitely coming."

RYAN KNEW HE HAD a good idea when he saw the response to the pizza delivery. A few minutes later, he revised his opinion upward when another delivery brought a half keg of beer. Two pizzas disappeared quickly. When the group started on a third, Ryan decided two pieces were enough for him and he could talk while the others ate.

He glanced around the table. Professor Moreno nibbled on her pizza, looking uncertain about why she was there. She probably didn't need to be, but she was Camila's thesis advisor, and she deserved to be included. He cleared his throat. "You all know what we're trying to accomplish beyond getting Camila her PhD."

"Mine too," Rafael Rodrigues, their materials engineer, said.

Ryan nodded and continued. "There's a danger that an existing company might see what we're doing and start their own research into the use of the Stenhouse Field. I've been talking to attorneys about what we can do about that."

"Can't we just patent Camila's design?" Ben suggested.

"According to the patent experts I've consulted, it's too early for a patent application. When we have a working generator prototype, perhaps we can get a patent. Unless someone else has already done so, of course."

"If someone else gets the patent, your company will be out of luck," Rafael said.

"It would depend on how much the patent claimed, but basically, you're right. In fact, we could already be out of luck if Stenhouse had applied for a patent for his field generator design, but he only published the design and left it in the public domain so that other scientists could investigate the field. The attorneys suggested we do something similar, an alternative called a defensive publication that would preserve our ability to exploit Camila's work."

Rafael looked puzzled. "How is that different from a patent?"

"In short, it establishes prior art, putting the work in the public domain. We could do that with little more than a drawing of the generator, assuming it went beyond what Stenhouse published."

Rafael scratched the back of his neck. "How does that help?"

"Others could work on developing the technology but couldn't patent their work, so they wouldn't be able to stop us from building a generator using Stenhouse Fields."

Rafael wasn't satisfied. "But a bigger company could beat you to putting a product on the market."

Ryan nodded. "Yes, that's a risk. But another company might be reluctant to use a design they couldn't patent and, therefore, control."

"Is that likely?" Camila said.

Ryan shrugged. "Less risky than hoping no one else patents a design. I wanted to get input from everyone involved, but we—and I mean the potential owners of a future company—need to come to a decision. Not tonight, but the sooner the better."

"While you're thinking about that, you should consider formalizing the company you're talking about creating," Mike said. "It's not too soon to look into investors to finance what happens after Camila graduates."

"That brings up something else." Ryan looked at Ben. "I assume you've been looking for a position now that you have your PhD. Anything to report?"

"I've interviewed at two companies, but neither has made an offer yet." Ben frowned. "One of them asked about my relationship with Camila. I got the impression they were interested in knowing more about her work."

Ryan nodded. "I can guess what company you're talking about. I have a different idea I would like you to consider. Come work for me until we have a company."

"I know nothing about your business. Why?"

"I'm an investor, Ben, and what we're doing could be my greatest investment yet. I'm going to need a technical consultant who can advise me intelligently on issues as they come up." Ryan grinned. "This probably doesn't mean much to you, but I'll match anything one of those other companies can pay you." He looked at Gabby. "Then again, now that you've started a family, maybe it is important to you."

"Let me think about it."

J ÜRGEN HAD HOPED THE meeting would distract him from his worries, but Ryan's information about patents only gave him more concerns. He had a copyright on V-World's code for *A World of Your Own,* but it only covered the interface between the artificial intelligence generator and the graphics, with a few minor AI improvements. Most of the program was based on licensed software from other companies. He had exclusive agreements with them, but that wouldn't prevent someone from using software developed by competitors of his suppliers or developing their own.

He had relied on being first, a belief reinforced by an offer from another game creator to buy his company. The amount offered had been mind-boggling, but he cared little about money. He owned a modest home outside Schenectady and devoted most of his time to V-World. Almost everything the company earned was invested back into his product.

He tried to tell himself he was being paranoid, but the visit from the PPC agent was never far from his mind. If the government came after him, because he was gay or for any other reason, he could lose it all.

Camila and her friends impressed him. And Gabby's involvement made it a family matter. He could sell V-World and invest in their company, solving his problem and supplying at least some of the money they would need. He shook his head. *I'm being ridiculous.* But he couldn't shake the thought.

"Getting money for new long-shot investments is still tough," Ryan told Mike. They sat in Ryan's office in Albany after Ryan finished another frustrating phone call. "I'm talking to manufacturing companies with no real skin in the game in information technology, but everyone is too scared to commit to anything."

"So what are we going to do? We could try the government again."

Ryan grimaced. "I've put out a few feelers, but right now, all we have is a grad student trying to get a PhD. We need more."

"Like an actual company?"

"Which we can't do without more financing. I can't handle it myself."

Mike shook his head. "And we can't get more financing without a company to finance."

"Exactly. Nobody wants to take a chance in this environment, even after two years. The politicians keep saying that it will all get better soon, but the government isn't doing much to make it better. It's hard to convince investors how big this could be when you can't speak the language. If we could get Ben to come aboard, he could help."

Ryan stared into space for several seconds. "We need that defensive publication to protect us. Ben could deal with it if he came aboard. I would rather not distract Camila with it."

"I'll call him and put some pressure on him."

"Make it fast. If he gets an offer and accepts it, getting his time will be even harder."

Ben had applied to several places for work, but hadn't pushed it as hard as he should have. Gabby worked for her brother, so he had concentrated on opportunities in the Capitol District. The college had offered a post-doctoral position on a research team, but it came with a three-year commitment. Other possibilities would require relocation or also came with implied commitments.

Camila would finish her thesis well before that, and she and Ryan would want to start their company then, if not earlier. Ben knew Ryan was looking for financing, and Mike had called that morning, renewing Ryan's offer of a technical consultant position.

This wasn't what I envisioned when I got my degree. He sighed, thinking about how his life had turned out. A simple agreement to help Camila, admittedly in part because of her attractiveness, had set off a chain of events that pulled him in. He wasn't complaining; if he hadn't gotten involved with Camila and her brainchild, he wouldn't have met Gabby. That meant more to him than the possibility that Camila might make him rich someday. But it wasn't the career he had planned.

He told Mike he would give him an answer within a day or two. But first he had to talk to his wife.

Gabby returned late that afternoon. She looked a little frazzled, probably as much by the commute from Schenectady as her workday, but she kissed him enthusiastically when he greeted her at their apartment door.

When the kiss ended, she leaned her head back and looked at him. "You usually react more when I kiss you like that. You're not tiring of me already, are you?"

Ben forced a smile and kissed her again. "Mike called to put some more pressure on me to come work for them. I promised to decide after I talked to you."

"Ryan would match anything anyone else would pay." Gabby released him. They moved into the tiny living room and sat down on the couch. "I understand why you don't want to be a consultant, but it won't matter in the long run if the plans work out."

"I suppose. And I've been putting off accepting any other offer so I would be available then, but I can't sit around until Camila finishes."

Gabby nodded and waved her hand around the apartment. "We've talked about starting a family. I haven't pressured you, but we should save for a house, and we can't do that on what Jürgen pays me alone."

Ben hugged her. "You've been more understanding than I deserve, and I love you for that." He paused. "Among other reasons. But you're right on all counts. I'll call Ryan tomorrow and tell him I'm in."

"Are you sure?"

He held her a little tighter. "Absolutely. Then we can start making babies."

Gabby kissed him. "I do like the way you're thinking."

B EN ACCEPTED RYAN'S OFFER the next day. When he reported to Ryan's office
suite in Albany the following Monday, Dani met him and brought him to the
office Ryan had assigned him.

The room was much better than the teaching assistant cubicle he had used at
the college. A large built-in desk with an expensive chair filled one corner. Two more
comfortable chairs provided seating for visitors. A large monitor sat on the desk,
hinting at the state-of-the-art workstation hidden in the bowels of the desk. One
wall had a counter with a coffee machine and several cups. A couch on the opposite
wall offered additional seating. Ben was impressed.

"Ryan believes that a comfortable workspace improves productivity," Dani
said. "Mike and I have almost identical offices."

A door next to the coffee counter opened onto a small closet. Dani took his
jacket and hung it up. "You'll find documents needing review on your computer,
but Ryan wants to speak to you right away." She grinned. "I hope you didn't think
you'd be having it easy here."

"I was more afraid I would." He followed Dani out into the corridor and down
to another office. Dani knocked on the door, and he heard Ryan invite them in.

Dani had said she had a similar office, and Ben assumed Ryan's office would be
larger. But it was the same size and furnished the same, except for some decoration,
mostly pictures.

Ryan rose from behind the desk and shook his hand. "Welcome, Ben." He
paused. "I'm afraid you won't have much time to get used to working here. You'll
have to hit the ground running."

"I'm not sure what I'm even here to do."

Ryan smiled. "Oh, you're going to be very busy. For starters, you probably
already expected to act as a liaison between Camila and us." He paused as Ben
nodded. "That's important, of course, but I need you for a lot more. In the short
term, I think you would be the best person to work with the lawyers for this
defensive publication they've suggested. Also, I will need you to provide technical
expertise when I talk to prospective investors. My ignorance in that area hasn't
helped our efforts so far.

"In the longer term, that podcast McGavock put out last month gave me an idea. Our position would be strengthened if we publicized what we're doing." Ryan clapped Ben's shoulder. "That will give you something to do in your spare time."

Ben shook his head and looked at Dani. "I guess my fear was unfounded."

Dani grinned. "You got that right."

32

C AMILA TOLD BEN WHEN the prototype design was ready, shortly after New Years. Ryan suggested a celebration and reserved a side room at their usual pizza restaurant so they could party without disturbing or being disturbed by the other patrons.

Ben had made progress, too. With Camila's final design completed, he could fill in the remaining details on Camila's work, fulfilling the legal requirements to prevent someone else from patenting a similar design. He also published a short article about the project in a college magazine. Two local news sources picked it up and a local broadcaster produced a short segment.

"So, can you explain what you have in words we business types can understand?" Ryan asked Camila when the pizza consumption slowed.

"OK, I think so. The object of the prototype is just to show that we can release energy this way. We want the output to be small, and expect it to be less than the input for the first test."

Mike swallowed a bite of pizza. "Why?"

"Because we don't want to blow up the building." Camila smiled. "This is just a proof of concept, not of any more use than that. To make the device simpler, it will use krypton, a non-reactive gas, as fuel. It will blow tiny quantities of the gas into the Stenhouse Field and, we hope, convert at least ten percent of the fuel's mass to energy."

"That's still more energy per gram than what nuclear power plants produce," Jürgen said.

Camila nodded. "Yes, but we'll be converting a tiny amount of matter. We chose krypton because its molecules are more massive than iron, but not by that much."

"The process won't work for elements less massive than iron because they take more energy to pull the molecules apart than is released as a result," Ben added.

"I thought nuclear power plants use hydrogen," Dani said.

"Fusion plants, not fission generators," Ben answered. "Fusion generates power by pushing atoms together, not pulling them apart."

"The first prototype will also test a couple of my candidates for materials that resist neutron damage," Rafael said. "A Stenhouse Field may adversely affect materials currently used, and we'll need data on that."

"Basically, the idea is to break nuclear bonds and release protons and neutrons, right?" Dani asked.

"Yes."

"Doesn't that change the fuel into some other element?"

"Removing protons does," Camila answered. "That brings up an interesting possibility. If we could control the number of protons removed, we could change a heavy element like lead into a lighter element. Gold, silver, platinum. Anything with an atomic number between 82 and 26."

Dani smiled. "Really? Gold?"

"Theoretically. I don't know how we could control the reaction that precisely, though."

"Oh." Dani's smile faded.

Ryan shook his head. "I think I understood about half of that. But thanks for trying."

"The Mechanical Engineering Department has a workshop staffed by students," Professor Moreno said. "They should be able to fabricate some of the parts for your first prototype. I can have them check out the design and determine how much they can handle."

"What about what they can't handle?" Camila asked.

"We'll figure that out after they've looked at it. Having a mechanical engineer evaluate the design is a prudent idea, too."

Camila nodded. "I was concerned about that. Designing something like this isn't what I trained for."

B EN AND CAMILA BROUGHT the design to the workshop together so that the students would know them both. Although several of the students working there would contribute to the construction of the generator, Ben had worked with Martina

Patel, a senior studying mechanical engineering. She had several commendations for the quality of her work, according to Professor Moreno, and she would lead the work.

"This sounds like an interesting project." Martina looked at both but rested large brown eyes on Ben. "Do you think the concept will work?"

"We're hoping," Ben answered. "Let us show you what we need."

"Sure. We can use this terminal here."

Martina logged onto the workstation and helped Camila download the design from her reader. A few quick commands brought the design up on the monitor.

"We've tried to keep the initial design as simple as possible," Camila said. "Gas is injected into the reaction chamber here. The power output will be low enough so that neutron radiation won't be a problem, but we'll need precise measurement of the gas flow rate."

"What are these slots in the reactor chamber for?"

"That's where I'll attach the electronics needed to produce the Stenhouse Field. The holes are for the field emitters."

Martina nodded. "What about the byproducts? As I understand this, you'll get everything from krypton down to nickel or iron. That includes arsenic and maybe a couple of other dangerous elements."

"Only trace amounts," Ben said. "That will be a problem eventually, but then we'll keep everything in the reaction chamber until it's completely converted. At that point, we should be able to use magnetism to move the byproducts out."

Martina smiled and shook her head. Her long black hair swept along her shoulders as she turned back to Ben. "If one of you could work with me, I can look at the design and get started right away."

"I can help you," Ben said. Getting out of the office for a day and working on the design appealed to him, even if it was more engineering than physics.

33

S ENATOR CORREIA WAS BACK on *Washington Views* with Kendra McDermott. With the election only ten months away, his favorability rating with his constituents seemed better, although it was hard to tell without good polls. Still, his instincts told him it wasn't enough, and he could still lose his seat in the Senate. He hoped another interview with Kendra would improve his position.

After a brief introduction, Kendra got down to business. "Most political pundits believe your reelection is far from certain. Your opponent, Allison Hayden, has been particularly critical over your stance on gay rights."

Correia smiled. "Of course she has. That's politics. But most of my constituents agree with me that unproductive members of society should contribute in other ways. I'm not against gays or gay marriage, but you must admit there is something less significant about a union that doesn't provide much-needed new members to our nation."

"Some of your constituents have agreed with violence," Kendra said.

"You're referring to the demonstration in Rochester two days ago?"

"Given the injuries and property damage that occurred, one might call it more a riot than a demonstration."

"Liberal elements confronted legitimate protesters. The result was unfortunate, but doesn't change the facts. I have no wish to restrict the rights of any citizen, but we all have responsibilities too. If one group chooses not to acknowledge their duty to the United States, then the United States should force them to help in ways that do not infringe on their rights."

"Such as with higher taxes?"

Correia nodded forcefully. "Exactly. At least for starters."

"For starters?"

"I haven't proposed further measures as yet. Circumstances could require something more in the future."

Jürgen found Senator Correia's past speeches unsettling, but Correia had appeared on a couple of talk shows, always with the same complaints, and nothing seemed to come from it. He heard nothing further from the Privacy Protection Commission and his concerns faded. The Rochester incident was frightening, and he hoped it would bring some sanity to the issue, but Senator Correia's interview had dampened his optimism.

Then Paige, his sometime receptionist, sometime tester, showed him a recording of an interview with Texas Senator Valarie Peterson the day after Correia's. The interviewer, Kendra McDermott again, asked Peterson about her support for Correia, and she had admitted agreeing with his statements. But then she had taken it further.

"These people have all the benefits of our society, financing their unproductive lifestyles with government handouts, owning businesses dependent on the infrastructure we all pay for, and all the other perks of living in a thriving, open democracy like ours. We must have laws that will make them contribute."

Some would write off such talk as political posturing that ran counter to the facts, something no one with any sense would pay attention to, but Jürgen had been a young child in the 2020s and remembered how much trouble similar rhetoric had caused. His parents and their attitudes, he felt, were a product of that time. The divisiveness from that period had never completely died away; it seemed possible that it might be renewed.

"You look worried," Paige said. "I was afraid you might be."

Jürgen nodded. He hadn't told Paige he was gay, but she figured it out, anyway.

"You're still dating that physics student, right? This shouldn't affect you."

"We're just friends."

"Nobody knows but family and friends." Paige smiled.

Jürgen wasn't convinced, but didn't want to take any drastic actions solely on his fears. There were other people involved, too, and he needed their feedback.

"Let's schedule a company meeting for this afternoon before everyone leaves," he told Paige.

"Diego too?"

"It's two hours earlier in Texas. He might be working, but if he's available, sure."

"I'VE HAD THIS OFFER for a while," Jürgen told the people seated around him. Diego Lopez listened from his home in Texas. "I didn't think I would ever accept it, but recently I've been rethinking that."

"You're selling the company?" Gianna Coleman said. "What will happen to us?"

"I am thinking seriously about selling, but I wanted to talk to you before I decided. The buyer made keeping the current staff part of their proposal. In fact, I will probably stick around for at least a while to manage this operation, so not much will really change."

"Can I ask why you're selling?" Chuck Sanchez asked.

Jürgen frowned. "I don't want to go into details, but I want to cash out and use the money for another investment."

Chuck and Gianna looked puzzled, but Paige nodded with a smile. On the workstation screen, Diego's smile was more a smirk.

"When will this happen?" Chuck asked.

"I don't know that it will happen. I have to talk to some other people first."

RYAN TOOK THE CALL immediately when his secretary announced Jürgen's call.

"I thought about talking to Camila," Jürgen began. "But finances are your part in this, so I'm talking to you first."

"Sure," Ryan answered. "What do you want to talk about?"

"I would like to invest in the company you are forming with Camila and the others."

He once said he had all his money tied up in his company. "OK. How much are we talking about?"

"I'm not sure how much I'll have to pay in taxes, but in round numbers, about fifty million dollars."

The mention of taxes was a clue. "You're selling part of your company."

Jürgen nodded. "I have a very nice offer for all of it."

Ryan scratched his head. "I don't understand. Is this because of Camila?"

"Not entirely. I have reasons for wanting to sell that have nothing to do with her project."

"But risking it all at this stage? Are you sure you want to do that?"

"I have modest needs. I would keep a few million, enough to live very comfortably and securely."

Ryan stared at the phone for several seconds, trying to wrap his mind around what Jürgen was saying. "I assume you would want a piece of the company?"

"Would ten percent be too much?"

Ten percent was less than a quarter of what they had planned to offer to investors. The proposal was a great deal for the company and could be leveraged to get favorable terms with other potential investors.

"The others would have to agree, of course," Ryan said. "Let's get together on Friday night to discuss your proposal."

SOMETHING WAS IN THE air as Camila and the others took seats at a table in their usual pizzeria. Ben and Gabby paid more attention to each other than the group. To a lesser extent, that was true of Mike and Dani. But Ryan and Jürgen had an expectant look, as if they knew something was about to happen.

Ryan waited until they ordered pizzas and beer before he leaned toward them. "Jürgen has a proposal." His voice was low but urgent, and everyone else leaned in to hear him better. "He would like to invest approximately fifty million dollars for ten percent of the company we are forming."

Camila stared at Jürgen, but he only smiled at her. She wanted to say something, but her surprise made her speechless.

Mike stepped in. "I thought all your money was tied up in expanding your company."

Jürgen nodded. "It is, but if you accept my proposal, I will sell V-World." He stared at Camila. "I already have a buyer and we've reached a tentative agreement."

"Why?" Ben asked. "Is this about Camila?"

"I have reasons for wanting to sell V-World. This project . . ."

But Ben interrupted. "What reasons?"

"What difference does it make?" Dani said. Her voice was sharp, and Camila wondered why. "You need investors and now you have one."

Ben looked at her, eyebrows raised, but didn't respond. After a few seconds, he turned back to Jürgen.

"I believe in what you're doing," Jürgen said.

Ben was still frowning, but Ryan sat back and smiled. "Maybe we should talk about this again after we've had time to think about it. We can decide whether to accept Jürgen's proposal at next week's meeting."

Several faces showed reluctance, but there were nods, too. The conversation turned to Camila's pending prototype test.

Martina Patel had delivered the parts for the reactor chamber as promised a week and a half before, but that was the more straightforward part of the prototype design. Camila had to build the electronics to generate the field, including the emitters that would extend into the reaction chamber and control the shape of the field. She had Stenhouse's work to guide her, but the work had progressed slowly.

The others encouraged her, and she promised she would try to have better news in a week. When they stood up to leave an hour later, Camila hugged Jürgen and whispered, "Thank you," in his ear. "Stay a little longer. I want to talk."

J ÜRGEN WAS GETTING HIS coat, ready to go home, but he frowned and motioned to a chair. Camila didn't take her eyes off him as she sat down and watched him pull up a chair next to her.

Camila wrinkled her mouth. "Jürgen, I appreciate it, but why are you doing this?"

"Everything I said was true."

"Except you didn't tell us why you want to sell your company."

"I think Dani figured that out. Because it's dangerous for someone like me to have the attention V-World gets me. Besides, I wanted to help you."

Jürgen regarded Camila as she nodded and smiled. He was as much a victim of that smile as any man, despite the lack of sexual attraction. Her smile was a window into all the warmth and sparkling optimism he had learned she possessed.

"I can almost understand the effect you have on normal males," he said. "I value your friendship more than you know."

The smile disappeared, replaced by an angry glare. "Don't say that. Never say that."

The sudden change startled Jürgen. "What? That we're friends? I thought we were."

Camila jerked her head. "No, never say that other men are normal and imply that you aren't." She grabbed his hands. "I value your friendship too, but you're as normal as anyone."

Jürgen knew what effect he had on women. He saw it in Dani and Martina in their group, and Paige at his office. Sometimes he even saw it in Gabby. But, while he was sure

Camila was sincere about their friendship, he didn't see it in her. He wondered if she realized the implication.

"WHAT WAS THAT ABOUT?" Camila asked Dani when they got back to their apartment. "You seemed to be mad at Ben."

Dani shook her head. "I'm sorry about that. But Jürgen obviously didn't want to go into his reasons for selling, and Ben was being nosy. I think I know why Jürgen wants to sell, and he wouldn't want to admit it to Ben."

"Why?"

"He wants to keep a low profile, given the trouble about homosexuals lately. Being the owner of V-World draws too much attention."

Camila stared at her, and Dani smiled. "Yes, I know Jürgen is gay."

"How long have you known?"

"Since the first time you went out with him and came home saying you were just friends. If I wasn't sure then, that hug between you tonight cinched it." Dani looked Camila in the eye and held it. "That was definitely a hug between two friends and only friends. For both of you because you're both gay."

"I'm not. I mean, I don't know. I . . ." Camila clamped her mouth shut, unsure of what to say. *Is Dani right?*

ACCEPTANCE OF JÜRGEN'S PROPOSAL was never in doubt. Camila and Ryan didn't have to try hard to convince Ben and Gabby that a known investor like Jürgen was better than getting involved with someone they didn't know. It would take time for Jürgen to complete the sale, and Camila didn't need the funds yet, but Jürgen was formally admitted to the voting group members at the meeting a week later.

Camila's progress report was less encouraging. "The control circuits I made using Stenhouse's design work, but the emitters are giving me trouble."

"What's the problem?" Ben asked.

"Stenhouse could create the field and investigate it, but he wasn't concerned with controlling the shape of the field. It wasn't important to him, but it will affect our reaction. The emitter shape seems to be critical, but I haven't been able to come up with the result I need. It's all extremely sensitive."

"How can we help?" Jürgen asked.

"I'm just winging it right now, but the tiniest change in the emitters changes the field drastically." Camila looked at Gabby. "Maybe if I could simulate different shapes accurately, I might get further. I'm having trouble adapting your algorithm to the problem and my time on the quantum computer is limited."

Jürgen smiled. "I have an employee who is an expert on writing software for complex problems. I could loan her to you."

Camila returned the smile. "Do you think she would stop working on games for a while?"

"Well, she works for me and she's my baby sister, so I think I can get her to do it."

Gabby shook her head. "Of course, I'll help." She glanced at her brother. "It would be nice to do something useful for a change."

Jürgen groaned and put a hand over his heart. "Ow. That hurts. My own sister."

$\mathcal{C}$AMILA, WITH HELP FROM Gabby, finished a program that could simulate the complex field interactions Camila had been dealing with. That increased the accuracy of the simulation enough to allow Camila to complete the emitter design. Gabby and Jürgen helped create software to control the Stenhouse Field. Early in March, Camila met with Gabby and Ben to review their status.

"I'll program the FPGAs this weekend." Gabby referred to the Field Programmable Gate Array components that would execute the controller software they had written.

"Good. I'm going to go through the test plan one more time. I want to double-check the expected energy output calculations versus the flow rate. We don't want to make any mistakes."

"Not with the output being so sensitive to flow." Ben leaned toward Camila. "You look worried. You've been over that several times already. This will work."

Camila tried to smile. "I'm not worried about the test. I'm very grateful for the work you've all done to support me."

Ben grinned. "No problem. You're going to make us all rich."

"That's what I'm worried about. When we decided to forego seeking a patent, it seemed like the right move, but somebody else could still develop their own generator and take over the market before we have a product."

Ben nodded. "There is a danger. We made the basic design of the generator public knowledge when we did the defensive publication. That, and proof that the design works, could be enough for some large energy company. But I think we would hear about it if somebody were using your research."

"You're probably right."

"So, when are we going to test the prototype?" Gabby asked. "I know we have to be careful, but we don't want to give the competition any more time than necessary."

This time, Camila's smile was genuine. "With some help from Martina, I should be able to build the emitters in a day or so. With them and the controller ready, I think we should schedule a test for Wednesday night and invite everyone to attend."

T HE SCHOOL HAD AN offsite laboratory built to host potentially dangerous experiments. On the afternoon of the test, Camila, Ben, and Martina installed the prototype in a basement room, setting up instrumentation to monitor the prototype's energy output, measure neutron generation, and record other relevant data.

That night, everyone met in a conference room in the laboratory. Professor Moreno came, but the entire Physics Department knew about it, so Camila wasn't surprised to see Professor McGavock as well. She was startled, however, when Doctor Stenhouse arrived with him.

"This is a big moment for you," McGavock said. "If the test is successful, I'll do another podcast and get you some publicity."

Camila smiled. "You don't have to go to all that trouble." She turned her attention to Doctor Stenhouse. "And you came all the way from Spokane? We appreciate your support for our efforts."

"It will be a pleasure to see my work put to a practical use," Stenhouse replied. "It was the hope of something like this that led me to leave my field generator available and not patent it. I'm so happy that you have done the same with your design for extracting energy from the Stenhouse Field."

"Thank you, Doctor. I appreciate your support."

Stenhouse smiled. "You're obviously very talented as well as beautiful. The idea of using the Stenhouse Field to release energy is brilliant!"

Camila could feel her face warming. She looked at Ben, standing nearby. "That was Ben's idea, I'm afraid. I suggested using it for space travel, and Ben thought clean energy was more important."

Stenhouse's eyebrows shot up. "Was it his idea to inject a controlled flow of krypton gas into the field?"

"No, that was mine. That will be the basis for my thesis, of course."

Stenhouse took her hand. "Then you're the one who took my research and showed how to exploit it. You're being generous, trying to give the credit to your associate, but I stand by my statement. Ben has made a name for himself already by showing I was right."

B EN WAS FINDING IT difficult to stand next to Camila and Stenhouse without reacting. Stenhouse was obviously flattering Camila, and, from the way she was blushing, Camila was eating it up. Ryan had warned her about Stenhouse the previous year. Then, Stenhouse had been taciturn, but now he tried to be charming.

Camila could certainly inspire a man to attempt charm, but Ben doubted that was the reason for the change in Stenhouse. *I was worried about other companies, but McGavock and Stenhouse could be a more serious threat.*

He looked around the group. Ryan and Professor Moreno were frowning, perhaps thinking similar thoughts. Jürgen, however, seemed amused by the exchange.

"Everything should be about ready." Ben turned to Professor Moreno. "I suggest we go down to the basement. Professor, could you show them where to go? I need to talk to Camila for a second."

Moreno nodded and gestured for the others to follow her. When they were gone, Ben turned to Camila. "Stenhouse is being rather pleasant this time."

Camila chuckled. "Yes, it's almost embarrassing." She paused and stared at him. "Don't worry. I still don't trust him."

Ben looked at her for several seconds before nodding. "I should have known he wasn't fooling you."

"Yes, you should have." She wagged a finger at him. She wasn't smiling, but he could see amusement in her brown eyes. "I'm certainly going to be careful with Professor McGavock and anyone associated with him."

"Of course." He laid a hand on her shoulder. "So let's go destroy some krypton nuclei."

36

C AMILA STOOD AT A control console to one side of an observation window. On the other side of the glass, the prototype was almost invisible, buried in shielding and sensors, but everyone crowded around, trying to see.

Camila's primary goal was to show she could use the Stenhouse Field to generate energy, but she also wanted to confirm the simulation predictions. After months of study, she understood the expected reactions, but the possibility of unforeseen phenomena was significant.

"I'll start at one microgram of krypton per second," she told the watchers. "I'll increase the pressure, and therefore the flow rate, as the test runs."

Camila turned back to the console and flipped a switch. "Here we go." She ran her eyes over the bank of instrument displays. "About twenty watts." She paused and watched the display, trying to ignore the excited comments from the observers. "Two micrograms per second and forty-five watts."

"It seems more efficient at the higher flow rate," Gabby said.

"I expected that. The atoms from the initial flow have become lighter elements, but the lighter elements will continue to react and add to energy from the added krypton until the atoms get down to stable elements like nickel and iron."

"The reaction chamber is made of stainless steel because that doesn't react with the field," Ben said.

"Isn't that about what you would need to power a light bulb?" Mike asked.

Ben nodded. "Using a millionth of a gram of krypton."

"So a gram could power a million light bulbs?"

Camila kept her eyes on the instruments as she answered. "Theoretically. This unit isn't capable of handling that much power, though. It would probably blow up."

She frowned. "More radioactivity than I expected. Some isotopes are lasting long enough to decay."

Gabby blinked. "I thought only heavier elements were radioactive."

"No, any element can be radioactive if the number of neutrons varies too much from the number of protons," Ben explained.

Gabby looked through the window while backing away. "Is it dangerous?"

Ben put an arm around her. "Not at this power level. Nothing significant is getting out of the test room."

"Neutron energy is less than I expected, though," Camila said. "The quantity fits estimates, but they don't have as much energy as we expected."

"Did your calculations take gravity into account?" Stenhouse asked.

Camila turned and stared at him. "We assumed any gravitational effects within the field would be negligible."

Stenhouse shook his head. "You're warping space and that affects gravity, but how much depends on the configuration and strength of the field. Gravitational effects could be reduced or increased, depending on the location relative to the field as well. My experiments didn't deal with those issues." He frowned. "My detractors used that to question my results until Ben proved them wrong."

"This is just a prototype," Ryan said. "We'll be making improvements. This is still a successful test, though. Right?"

Camila smiled. "Good enough to finish my thesis."

Professor Moreno was silent during the test, although her attention never wavered. Now she smiled and hugged Camila. "Absolutely. Congratulations, Camila. You and your team have made an incredible breakthrough."

After that, there was a flurry of handshakes and shoulder clapping. Stenhouse smiled at everyone and stood a little aside, but McGavock was in the center of the celebration, and Camila wasn't sure whether he was congratulating them or drawing attention to himself.

Camila, Ben, Gabby, and Rafael sat in Camila's lab in the physics building. While the test had been a success, proving they could use the Stenhouse Field to extract energy from nuclear fission, there were issues to deal with as they went forward.

"As we expected, debris from the krypton fission has gotten into the control mechanisms," Rafael said.

"I assume the debris is from elements lighter than krypton but heavier than iron," Ben said.

Rafael nodded. "Yes. That includes bromine and arsenic in small quantities, but mostly iron, cobalt, and nickel. That, at least, is good news, because it means the reaction is almost complete and therefore very efficient."

He paused and looked at the computer in front of him. "Some of these, especially the bromine, are reacting with the material they contact. The device will deteriorate rapidly because of this, especially when we increase the output and raise the operating temperature."

"Can we remove this debris before it reacts?" Gabby asked.

"If we improve the efficiency to get only iron debris, we could remove it magnetically," Camila said.

Rafael looked doubtful. "You'll need to get down to iron, or at least nickel, quickly. At operating temperatures, exposed surfaces could deteriorate fast."

"Maybe we could use a lighter gas mixed with the krypton to buffer the reaction," Camila suggested. "We would have to increase the flow rate to get the same output, though."

"It might need to be another inert gas like helium or argon," Rafael said. "Otherwise, the buffer could bond with the debris to complicate the reactions."

Camila nodded. "I'll think about it. Under operating temperatures, the atoms might not form bonds."

C AMILA MET WITH BEN, Jürgen, Martina, and Rafael two days later, after thinking about the problem and doing initial design work on a modified prototype generator.

"I don't think adding buffering gases will work," Camila said. "The reaction would be less efficient and the chemical reactions will add problems even with lighter inert gases."

"We have to make the fission go all the way to iron," Ben said. "Maybe a longer field would do that, even at an increased flow speed."

"Or you could recirculate the debris through the field until it completely fissioned into iron," Jürgen suggested.

Camila shook her head. "Either will require a new, more expensive prototype. It should be easy enough to add air to the krypton stream and get more data from the current prototype first." She looked at Martina.

The mechanical engineering student frowned. "Let me guess. You think I would be the logical person to fix the damage done to the prototype."

"Some of it," Camila agreed. "I'll need to build new emitters. The original ones are too damaged to give reliable data."

Martina nodded. "Rafael, if you could help me assess the damage, I'd be grateful. I'll probably have to build a new reaction chamber, but there may be other damaged components."

37

T HE TEST RESULTS CAMILA used for her thesis were available to the school faculty, especially to professors on the physics department thesis review committee. McGavock had full access to them within a week of the test. After familiarizing himself with the data and Camila's analysis, he sent the information to Stenhouse.

The next day, he called Stenhouse, who had returned to the Advanced Physics Institute in Spokane, Washington. After exchanging greetings, McGavock asked, "Have you reviewed the information I sent?"

"I have. Interesting. The idea of adding a light gas to the fuel could work. I assume they are continuing."

"Lopez will turn in her thesis for review soon. She has enough now and will probably end her ties to the college as soon as her degree is awarded. My understanding is that they will try to productize her work, and I won't have access to it then."

"I'm sure you're right." Stenhouse paused. "You sound like you have a suggestion to make."

"They're a small group and will have to grow substantially before they can develop a commercial product. There's an opportunity there for a company that can move faster. The bulk of the profits will go to the company that succeeds first in building a commercially viable product."

There was silence on the line, perhaps because Stenhouse was thinking, but McGavock plunged on. "We could get someone interested in supporting us. After all, you're the actual expert on the Stenhouse Field."

"I thought that might be where you were going with this. My employer is strictly a research facility, but I have some contacts that might be useful through them."

"I know people through the college as well," McGavock said. "So you and I are going to move on this?"

"It was my discovery. I should benefit from it."

"WE HAVE A PROBLEM," Camila told the group assembled in her office. "Injecting air into the stream is causing more performance degradation than we expected. More radioactivity, too."

She had passed out copies of the test analysis, and Ben was scanning through the pages. He looked up. "Traces of carbon, boron, and magnesium. The oxygen and nitrogen are fissioning too."

"Wouldn't that increase the output?" Gabby asked.

"For elements lighter than iron, fissioning absorbs energy," Camila answered. "The air is sucking up some of what the krypton is releasing."

Gabby nodded. "Of course. That's why fusion power plants use light elements like hydrogen and fission power plants use heavy elements like uranium."

Camila sighed. "The reaction chamber will have to be redesigned to be more efficient. We'll need a new prototype."

"You already have enough data for your thesis," Ben said.

When Camila responded with a nod, Ryan said, "Maybe it's time we spent some of Jürgen's money and set up shop away from the college."

"I hoped to get more data and improve my presentation to the thesis committee." Camila looked at Jürgen, who nodded. "But maybe you're right. Let me talk to Professor Moreno first. If she approves, I can finish up the paper and schedule the verbal presentation by late next week."

"We'll need a facility," Ryan said. "I'll investigate the possibilities while you finish. Ben, I'll need your help."

"Sure. Someplace local, I assume?"

"As close as possible. This is where we all live."

RYAN MET BEN LATER. "We're going to need office and lab space. For now, at least, we should find a good machine shop Camila can use for prototype fabrication."

Ben nodded. "Did you want me to look for places?"

Ryan shook his head. "The office and lab space are relatively easy, at least for our needs over the next few years. I found a machine shop in Schenectady that I want to check out, but I don't know anything about that kind of thing."

"Me neither. Maybe you could get Martina to help."

"That makes sense. I don't have her number."

Ben picked up his phone. "I do. I'm sending you her contact information."

Ryan's phone beeped, and he accepted the transmission from Ben's phone. There was a personal number and a number for the Engineering Construction Facility, and he tried the former first. Martina answered after a brief delay.

"I want to check out a machine shop in Schenectady," he told her. "After Camila graduates, we won't be able to use your shop for our work, and I wouldn't know anything about what I'm seeing. If you go over there with me, I'll pay you."

"You plan to build a company out of what you're doing, right?"

"Sure. Having our own machine shop is part of that."

"And you'll eventually be hiring people to work for you."

Ryan grinned. "I don't think I would run into any opposition from the others if I promised to hire you when you graduate."

He could see her smile over the phone. "Then there's no need to pay me. Will Saturday be OK?"

"Of course. I know you have classes and all that. Let me make arrangements with the owner and I'll call you back."

38

F ROM THE PARKING LOT, Welch Precision Machining didn't impress Ryan. The large building, painted a garish pink, seemed to give Martina giggles. Deep gouges in the asphalt surface rocked their vehicle as it parked next to the building. A dent in the wall looked like the result of a vehicle impact that hadn't been repaired, and he could see signs of rust along the edges of some of the corrugated metal sheets. Two trucks were already parked by the building.

They went through a scarred front door into a small reception area. The owner, Mary Welch, came in from an inside door and nodded to them. Loud noises came through the door, and she let it close before she spoke.

"Mr. Terry?" she asked.

Ryan stepped toward her and held out his hand. "Yes. Call me Ryan, and this is Martina Patel."

On the phone, Mary Welch had sounded subdued, showing little enthusiasm when Ryan told her he was interested in buying her business. Her appearance in person strengthened that impression. She looked like a woman in her fifties, although Ryan knew from his research that she was only forty-two. She was tall, but a little overweight, with the drawn face of a tired, defeated woman.

"Are you familiar with machine work?" she asked.

"Not at all." Ryan smiled. "That's why I brought Martina. She's a mechanical engineering student who works in a machine shop at the college."

Mary nodded, but didn't seem interested. "Well, let me give you the tour, then." She opened the door again and held it for Ryan and Martina.

Ryan looked around and knew his eyes had gone wide. The modest exterior and reception area hadn't prepared him for what he saw. He had expected old machines covered in grime and dirty oil, machines that had been used for a long time, but to Ryan's inexperienced eye, the shiny equipment seemed very well maintained. Martina was smiling as her head moved back and forth, trying to take it all in. She looked up at Ryan, and her grin widened as she nodded in approval.

"This used to be the most successful shop in the Tri-City area," Mary said. "We have numerical control, 3-D printing, and casting capabilities enough to make almost anything smaller than a tank, and can work iron, nickel, titanium—almost any material."

"Impressive," Ryan acknowledged. "But you're in bankruptcy."

Mary frowned and looked down. "We still get a fair amount of business, but the competition in this area is fierce. The company owes money on most of these machines, and business has declined because of the recession in the last few years."

Ryan gave her an encouraging smile. "Let's see what else you have here, and maybe we can come to an agreement that will help us both."

Mary didn't seem convinced.

R YAN ASSEMBLED EVERYONE AT their usual pizzeria, but, after meeting Mary Welch on Saturday, he called the meeting for Monday. "First, I want to hear from Camila," he told everyone. "What's the status on your thesis?"

"Professor Moreno has reviewed it and had a few minor suggestions that I've already fixed," Camila said. "I'm scheduled to go before the review committee next Tuesday. Professor Moreno isn't expecting any problems."

"Then your degree should be official next month?"

"Yes."

Ryan nodded. "And then your connection to the college will be gone. I've already gotten the ball rolling on creating a company. How does Energy Unlimited sound to everyone?"

Ryan scanned the group and saw only smiles and nods. "OK, it will take a while for all the formalities to be complete, but we need to move if we're going to get the machine shop. Otherwise, the bank will foreclose and any deal we could make would be worse."

"You're sure this is what we need?" Ben asked.

"It's incredible," Martina said. "It makes my shop look like a blacksmith's forge."

Ben shook his head. "If it's so great, why is it going bankrupt?"

"Ten years ago, it was the most successful in the area," Ryan answered. "After Peter Welch died, it faltered for a few years as his wife learned the business. It looks like her competitors took advantage and stole many of her customers. She was left with debt for all the machines and a heavy mortgage without enough loyal customers to keep up payments."

Ben frowned. "That sounds good, but how will we finance the purchase? We don't have Jürgen's investment yet."

"The deal is close to closing," Jürgen answered. "Perhaps before the end of the week."

Ryan shook his head. "We don't need it immediately. The business has $800,000 in debt that we would take on. I offered the owner a generous salary to manage the place and keep her current staff. Subject to the board's approval, of course. We can close with little initial expense, and it's a good deal for her as well as us. If the bank forecloses, she'll be left with nothing."

Ben frowned. "How big a staff are we talking about? Camila's needs won't be that great immediately."

"I've got some ideas for marketing that could bring back some of her old customers," Dani said. "We might even wind up making money."

"I think Mary Welch has taught herself how to manage a machine shop since her husband died," Ryan added. "I don't think she had any idea of how to handle competition, though. If we can market the business better, this will work."

"And I'll move in there?" Camila asked.

Ryan shook his head. "No, we'll have to find a separate place for offices and a laboratory. The machine shop doesn't have that kind of space and it's too noisy. It shouldn't be hard to find something not far from the machine shop."

T HE NEXT MORNING, JÜRGEN called Camila. "You looked a little stressed last night. Why don't I take you out to dinner tonight, away from Ryan and his plans?"

Camila hesitated, and Jürgen could see her frown. But her smile returned, and her shoulders moved in a shrug. "Sure, why not? You're right. All this is happening so fast that I'm getting dizzy."

"Martina told me there's a great Indian restaurant up Route 7 on the Vermont border. She said they make the best butter chicken."

"Butter chicken? That sounds like something that would be terrible for my figure."

Jürgen chuckled. "Fortunately, I like you for your mind."

"There's that. OK, it's a date."

"I'll pick you up at your office at five."

Camila agreed, and they ended the call. When Jürgen arrived that afternoon, she was staring at her workstation screen and barely acknowledged him at first. Finally, she shrugged, smiled up at him, and stood.

"Let's go before another thought grabs me," she said.

They walked out to the car Jürgen had waiting. After Jürgen told it where to go, they settled back for the ride.

"I thought the design for the new prototype was complete," Jürgen said. "You seemed to be pretty busy on your workstation."

Camila shrugged. "I was just checking out an idea I had."

"Anything good?"

Camila hesitated before answering. "I don't know yet. Probably not."

"Do you want to get my input?"

"It's pretty technical. I should talk to Ben."

Jürgen wanted to help, but had to accept that Camila had a technical problem that he wouldn't understand. Except for the people he worked with at V-World, Jürgen didn't have many friends. Women shied away when they found out he was gay, and straight men had trouble trusting him. He didn't have the courage to seek other gay men.

Camila was different, and probably the closest friend he had ever had. He hoped the others in the group would remain friendly, but they didn't know he was gay, and he was reluctant to test those relationships.

"Well, whenever you need help from me."

39

DANI BROKE THE PHONE connection and groaned. The wedding was less than two weeks away and the caterer was demanding to meet to discuss the reception, but her father-in-law-to-be wanted the advertising plan for the machine shop. It didn't help that the proposed campaign was her idea, but the timing was terrible.

Mike was no help. He seemed to feel that weddings were for the bride to manage, and the groom's role comprised little more than an occasional pat on the back. Her parents were even less helpful; once it was established that Ryan was happy to pay for the wedding, they limited themselves to encouraging words. They still lived in Ithaca, two hundred miles away, and both worked, making any more than that difficult.

At least Mike's mother, Ana Madison, had come to Troy the previous weekend to help, but then she returned to Arizona, so she couldn't help with the caterer. Mike could have helped with the marketing plan, but he was busy with Ryan arranging the financing for the debts that had almost bankrupted Mary Welch. She had to admit that was probably more important than her marketing task.

The concept that had gotten her into the situation was simple enough. First, they would tout Energy Unlimited as the wave of the future, a clean solution to solve the problem of supplying power to a country that depended so much on its availability. The campaign would try to position EU as the leader in exploiting the Stenhouse Field and relegate any possible competition to a mere imitator. That might even help in getting government grants.

The second part of the plan would be to advertise Welch Precision Machining as EU's preferred source for its manufacturing of advanced prototypes. There, the goal would be to lure old customers back and make the machine shop more attractive to new business. If it could become profitable again, the debt it carried wouldn't be a problem.

As always, the work was in the details. Advertising for the machine shop would be primarily local, but EU needed to be visible at least nationwide, if not worldwide. The tasks needed were significantly different.

By the time she went home, she had the outline of the local part of the campaign, but another conversation with the caterer had diluted any stress relief from her progress. Venting to Camila often helped at such times, but when she got back to the apartment they still shared, Camila wasn't home.

I T WAS ALMOST DARK when Camila finally got back to the apartment. Dani was stretched out on the couch, but she sat up slowly while Camila told her about her dinner with Jürgen.

Camila plopped down on the couch next to her. "So, how was your day?"

"Ugh!" Dani pulled her feet out from under her and onto the floor. "The caterer is giving me grief, and Ryan is pushing me on the marketing plan for Welch's." She stood and groaned. "Mike is working with Ryan on the financing and can't help with either."

Camila frowned in sympathy. "I don't know anything about marketing, but if there's something I can help with on the wedding planning. . .." She leaned back and sighed.

"You're about as energetic as I am." Dani sat back on the couch next to Camila and stared at her roommate. "Are you all right?"

Camila leaned back on the couch. "I don't know. I have the feeling I've led everyone into something over our heads."

"You've already proven you can generate power with a Stenhouse Field. That's pretty good for someone who is just out of school."

"And I'll get my degree thanks to that and all the support everyone has given me. But now what? Ryan wants to create a company to exploit my work, but I don't know if we can do that. We're so small, and we can't stop some bigger company from beating us to a product."

"Can they steal your idea?"

"We prevented anyone else from patenting my work, but not from using it."

"What are you going to do? Ryan is taking on a lot of debt for this."

Camila could hear the concern in Dani's voice. "That's what worries me most. I don't understand the money issues. How deep in is he?"

"I don't have all the details, but his biggest problem right now is the debt he's taking on for the machine shop. If we can make that profitable again, he'll be OK." She paused. "Let us worry about that. You need to worry about the technical stuff."

Camila nodded. "I have an idea that I want to discuss with Ben, but I really don't know if it will help. I could have talked to Jürgen about it tonight, but physics is not his strong point."

"YOU AND RYAN MIGHT be right about competition," Camila told Ben the next morning. "Why wouldn't they want to develop a Stenhouse reactor themselves if they could do it faster than we can."

Ben nodded. "We knew that was a possibility when we used that defensive publication to prevent someone beating us to a patent. It wasn't a permanent solution."

"We need to talk to the lawyers again, but I think I have an idea that might help. Instead of trying to improve our current design, maybe we can come up with a new design that we can patent."

"We designed our prototype for gas injection because it would be easy to control." Ben rubbed his chin. "Do you have another idea?"

"Maybe. The reaction that generates energy occurs at the boundary of the Stenhouse Field. We've created a cylindrical field and pumped krypton through it. Suppose we put the fuel source on the other side of the boundary?"

Ben's eyebrows shot up. "A solid fuel?"

"Sure. A hollow cylinder of lead or other heavy metal with the same cylindrical field inside it."

"Using a heavier fuel source would generate more energy, but controlling the reaction might be more difficult."

Camila smiled. "I didn't say it would be easy. And a lead core would reduce the need to shield from the radioactivity of intermediate isotopes. But if we can develop the idea far enough to patent it before we make it public, we have a chance at beating any competition."

"We would have to get a little luck. Someone could still beat us to market with the gas design."

"They might, but we could still make out all right if the new design is better."

"Our first design was public because the school had to know what you were doing for your thesis. We don't have that problem now." Ben paused. "But we would have to keep it secret."

"I know."

"We should talk to Ryan about this."

RYAN STARED AT THE computer screen displaying the status of his investments. The markets were recovering from the crash caused by the Newman-Fredrickson Privacy Act almost three years before, and he wasn't anxious about his own financial health. But the deal with Mary Welch would tie up a significant share of his more fluid assets, at least until they could tap into the proceeds from Jürgen's sale of V-World. He hoped to turn Welch Precision Machining into a profitable business, but that wasn't assured.

If Energy Unlimited failed, it would affect the others too, although the boost to their resumes would probably guarantee their ability to find positions elsewhere if they had to. He could give them some encouragement on that if they needed it.

But it was only a matter of time before someone decided to capitalize on Camila's work. Perhaps it had been a mistake to prevent patenting the method she had developed. That decision was never far from his thoughts since the successful prototype test.

Ben came into his office and quietly closed the door behind him. "Camila has an idea for a new reactor design." Ben spoke in a hushed tone, as if he didn't want to be overheard. "We need to run it by you."

"OK." Ryan smiled, hoping it would be encouraging. "How much will this cost?"

"We'll need another prototype," Ben said. "But that's not why we wanted to talk about it. As it stands now, someone will probably steal Camila's idea and push us out of business before we even get started. They can do that because we can't get a patent on the gas-fed design."

Ryan nodded and looked at him with more interest. Perhaps this would not be a hand-wringing meeting.

"Camila has an idea for an entirely original design. I don't want to bore you with the technical details, but we hope we can get a patent before any competition has the same idea. I think the new idea has a better chance to make EU successful, even if someone develops the original design first."

"How much work have you done on this? We'll need enough to base a patent on."

"That's the problem," Ben said. "It's just an idea right now. We have a lot of work to do on a prototype design we can build and test. I don't think we can get a patent until we've done that, and it will have to be done without possible competitors learning the details until we're ready."

Ryan nodded. "From now on, let's have our Friday meetings here. I have a conference room large enough, and I can have pizza delivered."

WHEN JÜRGEN'S SISTER MARRIED Ben, the event had been unpleasant until he met Diego and, through him, Camila. The attention he had received at the reception had been awkward, and he wasn't looking forward to a repeat at Mike and Dani's wedding.

But there was an obvious solution. "Why don't we go to the wedding together?" he asked Camila over the phone.

Camila hesitated, but then smiled. "That sounds like a great idea. Dani is pairing us in her wedding party anyway, and it'll be fun."

"You might get more attention from unattached males if I'm not with you."

"Is that supposed to be an advantage or a disadvantage?"

Jürgen smiled, but didn't answer the question. "Good, it's set then."

THE WEDDING AND RECEPTION that followed were held in a large hotel in Saratoga Springs. Camila sat with Jürgen, Ben, and Gabby at the wedding party table with the newlyweds. Ryan sat nearby with his ex-wife and Mike's mother, Ana. Sitting with them were Dani's parents and Martina and her date. The time before the meal was served was a time for socializing, though, and everyone wandered around the ballroom, greeting others and sampling from the appetizer tables.

After dinner, there were the usual wedding activities: cutting the cake, congratulatory speeches, and throwing the bouquet among them. Camila faked an effort to catch the bouquet and was relieved when one of Dani's cousins snagged it.

Later, there was dancing, and Camila learned Jürgen was an excellent dancer. He knew the latest steps and helped Camila master unfamiliar dances. She was a little out of breath when the tempo slowed, and Jürgen held out his hand. Others were already moving into the arms of their partners, and Camila shrugged and put a hand on Jürgen's shoulder. He placed a hand on her waist, and they moved across the floor in time with the music.

Camila felt comfortable in Jürgen's arms. He didn't hold her as close as some of the other couples, but that was understandable. She was fine with that, but she had expected to feel awkward and didn't. Her friendship with him was important to her, but that's all it was. Thinking about that aroused conflicting feelings, and she tried to distract herself by watching the others over Jürgen's shoulder.

Others were watching them, too, especially younger women. Curiosity or envy? She leaned toward Jürgen and whispered in his ear. "We're attracting attention from jealous women."

Jürgen chuckled and pulled her a little closer.

Camila almost resisted, but hesitated. "What are you doing?"

"We should give them more to be jealous about. I think that might benefit both of us."

Camila could understand his reasoning regarding himself, but why did he think she would benefit? Her uncertain feelings returned, but she rested her head against Jürgen's shoulder and stopped watching the others.

R YAN'S RELATIONSHIP WITH ANA post-divorce had been a series of ups and downs. They maintained a cordial but distant link for their son, but that had been threatened when Mike decided to join his father. Ana wasn't one to hold a grudge, though, and had long since forgiven him. Now, watching their son celebrate his marriage, Ana was in one of her friendly moods.

"They make a nice couple," Ana said.

Ryan had been looking away from Ana toward Mike and Dani wrapped tightly together on the dance floor. "Mike was only waiting for Dani to graduate. He's loved her for a long time."

Ana chuckled. "I was talking about their friends. Your physics genius and that tall hunk with her."

Ryan looked in the direction Ana was staring. "Camila and Jürgen? Yes, they've become very close."

"So is that going to be the next wedding?"

Ryan shrugged. "Maybe. I'm not sure how they feel about each other. I've seen no physical signs of affection, although they spend more time together than their work requires."

"She's gorgeous, and he's the best-looking man in the room. How can they not be together? Look at them!"

Ryan took a long look. They seemed to enjoy the slow dance, but somehow it wasn't as intimate as what he saw with Mike and Dani, or Ben and Gabby, who were also dancing. He didn't know how to answer Ana, so he just nodded.

41

T HE WEEK AFTER THE wedding, Professor McGavock did another podcast, with Arthur Stenhouse participating remotely.

"Arthur, you're working with a division of General Power to exploit your discovery of the Stenhouse Field," McGavock said.

"Well, we're still in the early stages of planning," Stenhouse answered. "The possibilities of generating clean energy from my work could be world changing."

McGavock nodded. "Indeed. The implications for climate control alone are incredible. But it was a PhD student here at my university that showed the technology could be used to generate power."

"Yes, you mean Camila Lopez. I've met her, and she is a very promising physicist. Unfortunately, she has chosen to exploit her work herself by starting her own company, rather than working with an established organization."

"Why unfortunately?"

"Something with so much potential needs a large support structure, and she won't have it, at least not in any reasonable time. The world needs Stenhouse energy as soon as possible, and I hope to deliver it much faster through my relationship with General Power."

"There are rumors that the German company Saubere Energieerzeugung might invest in Energy Unlimited," McGavock said.

"I doubt that will happen. I have also talked to leaders in the Department of Energy about government funding, and they have expressed reluctance to support an effort with foreign investors. Technology transfer is a problem, exacerbated by Energy Unlimited's failure to patent their generator design."

"And you wouldn't need foreign investors."

Stenhouse nodded. "Yes. I think we would be interested in bringing Miss Lopez into our project, but I don't see how she can succeed on her own."

C AMILA FROWNED AT THE display on her computer. Generating power with a Stenhouse Field within a heavy element core had seemed like a promising concept, but it was looking worse the more she examined the idea. She realized quickly that the temperatures created would vaporize lead and most other materials. Maintaining a stable energy output would be difficult.

As the field stripped protons and neutrons from the atoms of fuel, releasing energy, it would leave lighter elements that would mix with the fuel, contaminating it and making control even harder. She and Ben were considering a second gas-injection prototype to test the practicality of using a magnetic field to remove iron from the reaction chamber, but her calculations already indicated that wouldn't solve all the problems.

With the gas-injection design, they could manufacture a reaction chamber from nonfissile materials with high melting points, but that wouldn't work for a design where the lining was also the fuel. She needed to talk to Rafael Rodrigues, their Materials Engineer. He had just graduated and was joining them as a permanent employee, but he was with Martina, talking to Mary Welch about parts for a second prototype.

Camila was a physicist, not a mechanical engineer. She left a message with Martina and Rafael, requesting help.

P ROFESSOR McGAVOCK SAT IN the reception area for Senator Correia's office, torn between impatience and some nervousness about meeting the influential politician. Stenhouse should have been the one lobbying for government funding, but claimed he was talking to leaders on the west coast. The vice-president from General Power had been insistent that they both look for grants that might reduce General Power's outlay for the project.

The receptionist picked up a phone and listened for a second. Then she looked at him. "The senator will see you now."

He nodded and went through the door she opened for him. Correia looked up, but McGavock could read nothing in the politician's smile. Correia waved a hand at a chair next to his desk and McGavock sat down.

"You're here representing General Power?" Correia said.

"Yes. GP is beginning a development effort that could solve our energy problems. Government support would help speed a successful outcome."

"Your request mentions work by Arthur Stenhouse, a physicist in Washington. You say he is working with General Power."

"He has moved to Chicago to lead the effort. We plan to exploit his discovery of the Stenhouse Field to generate cleaner power with much greater efficiency than any current technology."

Correia frowned. "And you claim you have built a prototype that proves that the concept will work."

"A student at my college built a prototype. I am very familiar with her work and am working with Stenhouse and GP."

"This sounds familiar. I think someone else talked to me about funding a project based on this technology."

"That was the student. When the government declined to fund her work, she reached an agreement with a local investment firm. They need more funding to continue, though, and there are rumors they are negotiating with the German company, Saubere Energieerzeugung, to give them part ownership in the company she has formed to exploit this in return for support."

"This student is working with a foreign company? That could be a problem if this technology is as good as you say it is."

"That's one reason it would be better if General Power developed the first practical generator."

"So the student intends to develop this as well. If she built the prototype, won't she patent it?"

"Because her organization is so small, she knew somebody else might build a prototype and patent the design first, so she filed a defensive publication. That's a legal trick that prevents anyone, including herself, from getting a patent."

"But she has a head start on you."

"Yes, but a major company like GP can easily overcome her lead, especially if the government helps."

Correia nodded and asked more questions. After ten minutes, he leaned back in his chair. "This has been interesting, Professor. Thank you for bringing it to my attention, and I will see what I can do to help General Power."

42

B Y THE FOURTH OF July, Camila's PhD was official, Martina and Rafael had their Bachelor of Science degrees, and Mike and Dani were back from their honeymoon. The process to create Energy Unlimited was proceeding smoothly, and Ryan predicted they would have a company within two weeks. Ryan called for a celebration, and, after a brief discussion, the group decided on a picnic at nearby Thacher Park.

Camila tried but couldn't get the proposed new generator design out of her thoughts. When she mentioned it, she discovered she was not alone. After their lunch, she found herself in the center of almost the entire group. Seeing where the conversation was going, Ryan, Mike, and Dani disappeared with vague talk of checking out one of the hiking trails.

"We thought this would be a simpler design," Ben said when Camila showed him her work. "I'm not sure it's true."

"The lead will vaporize at operating temperatures," Camila answered. "Above three thousand degrees, everything between lead and iron will. Controlling the rate at which the lead reacts is tricky."

Martina shook her head. "That's not the way existing fission power plants do it. The fuel is in rods that limit how much fuel is available."

Camila frowned. "This is different, though. Isn't it?"

"Yes, but the principle still holds. Nuclear plants use cooling, too, to prevent melting of the fuel. Newer plants use liquid sodium, not water, so that's a consideration as well."

Camila nodded. "Of course. I knew that." She paused. "Maybe you can help me figure this out?"

"I think it might be over my head already. And Mary is looking forward to me working at the machine shop." Martina shrugged. "I think you need someone more experienced."

Martina was right. The first prototype had been a crude device, useful as a proof of concept, but only that. To move the generator design beyond it, their fledgling company would need people with skills far beyond Camila's, able to devote their full attention to Energy Unlimited. Martina would be finding her place at the machine shop and, in less than two months, starting work on her master's degree as well.

"We'll talk to Ryan and the others when they come back," Camila said.

AFTER JURGEN LEFT THE picnic, he told his car to take an indirect route home. That wasn't difficult; there were no direct roads between Thacher Park and Schenectady. Deep in thought, he paid little attention to the small towns and villages he passed.

At first, he reflected on the conversation after Ryan rejoined them. Thanks to him, they had the money now to hire people. He felt good about his contribution to Camila and the others. But that just brought his thoughts to Camila.

Could he say he loved her when he had no sexual desire for her? Love wasn't all about sex, and perhaps some people would tell him he looked at her like a sister. He had a sister, though, and there was an edge of competition in that relationship that didn't exist with Camila.

If he were honest with himself, he had to admit that the protection he received from being seen with her was an element of his attraction. The increasing hostility toward homosexuals was a constant concern, but who would suspect he was gay when he was with her?

That couldn't be all of it, though. Straight guys had close relationships with other straights. Why couldn't he have a nonsexual attraction to a woman? A woman who was, he was convinced, gay herself and could only have a similar platonic attachment to him.

He had an honest, open relationship with Camila. He had told her he was gay, something he had admitted to few outside his family. It would have been natural for her to tell him she was gay also, but she hadn't done so. At the wedding, she seemed to think that his fears about exposure didn't apply to her. He could only conclude that she hadn't admitted her sexual orientation, even to herself. Did Dani know, living with Camila for years?

One thought led to another. Now that Dani was married to Mike and living with him, Camila was alone in the apartment they had shared. She would probably want to find another roommate. That could cause complications.

Should he confront her? Would that help her find her place in life, or would she deny it and pull away from him? And if he made her realize she too was gay, what then? She would also understand the dangers of being gay in the current climate.

His car pulled into his garage, and as he entered the house, he had one last thought, in its own way, scarier than all the others.

C AMILA HADN'T SEEN PROFESSOR Moreno since the diploma presentation, but they had parted with assurances from Moreno that Camila could call on her if she needed advice. Camila didn't want to impose too much, so she made an appointment before returning to the Physics Building and her advisor's office. When Camila entered the office, Professor Moreno rose and came forward to embrace her.

"I hoped you would stay in contact," the teacher said. "I understand you are starting a company to continue your work."

Camila nodded. "We've gotten funding and are making plans. Moving past the prototype that I used in my research is beyond my expertise, though. We need to hire an experienced mechanical engineer to continue advancing. Eventually we'll need more, of course, but Ryan says that hiring too fast will complicate management issues when we need to concentrate on the technology." She paused. "I thought you might know someone who could help me, Professor."

Moreno smiled and waved her hand. "Please, Victoria. We don't have to be so formal now." She put a hand on Camila's arm. "And yes, I think I can help you. I assume you remember Doctor Curtis."

"Of course. He was on the board approving my thesis."

"He's in his eighties now. Still sharp, but he's been teaching for forty years and is finally retiring. I've heard he's looking for an opportunity to consult. I don't know if he would work full time, but I know he's an excellent engineer."

Camila rubbed her chin. "Maybe we don't need someone full time immediately. If he will work with us. . .. It wouldn't hurt to ask him."

43

Normally, Ryan would have researched Doctor Dean Curtis's career before offering him employment at Energy Unlimited. Restrictions on access to personal records enforced by the Newman-Fredrickson Privacy Act made that difficult. He had a few facts that Camila had passed on from her talk with Moreno, but the professor knew Curtis only professionally, and only slightly. Curtis had not been enthusiastic about Camila's thesis choice, but apparently approved the result.

Ryan made an appointment for Camila and himself to visit Curtis in his office at the college. When they arrived and Ryan introduced himself, Curtis greeted them neutrally. He scanned them curiously as he motioned them to seats.

Curtis looked at Camila and inclined his head slightly. "Congratulations on completing your degree."

Camila smiled. "Thank you, Professor."

"So, my curiosity is overwhelming. Why did you want to see me now?"

Ryan stepped in. "We've been told you're retiring from teaching but might be interested in doing some consulting as a mechanical engineer."

Curtis stared at Ryan and then Camilla. His lips curled slightly in what might have been a smile. "So it's true. You really are starting your own company to exploit your research."

"All the formalities should be complete within a week," Ryan said.

"I was pleased, and a little surprised, by your thesis," Curtis told Camila. "I thought it an engineering project, not suitable for a physicist, but you did a good job."

"But you were right. I proved my theory, but I don't have the skills to go beyond that. That's why we need help from an accomplished person like yourself."

Ryan could have sworn the old man was blushing, something he had thought impossible for a Black man. Apparently, Camila's charms worked on all ages.

"You're just flattering me. It won't work, young lady."

Ryan stifled a grin. *It's working very well.* "If you're interested, we would have to talk about your qualifications, of course. Professor Moreno recommended you, but we need to know more. Also, if we hire you, you'll have to sign an NDA."

"Why do you need a nondisclosure agreement? Everything Miss Lopez has done is public knowledge."

Ryan nodded. "You can talk about what Camila has done as much as you want. The NDA is for what she's planning to do."

Curtis smiled, a genuine one this time. "Of course. Then maybe we should talk."

T HE BUILDING WASN'T FANCY, but it appeared to be well-maintained. Camila walked into a small reception area, but they hadn't hired anyone yet to work the counter against one wall. She drifted toward the hallway at one end, taking it all in. More than ever before, it felt real; this was her company! Partly hers, anyway.

Ben entered the hallway from an adjoining room and smiled as he walked toward her. "Gabby will be in shortly. She's at the coffee shop in the next block, getting what she thinks we'll need for our first day."

"If it's a coffee shop, I'm guessing that means coffee and pastries, not office supplies."

"Actually, I think Ryan already had the place stocked with the usual supplies." He touched her arm. "Come, let me show you your office."

Ben took her down the hall, pointing out his workplace and a common office with desks for people like Ryan, Martina, and Rafael who would usually be elsewhere. "Mike, Dani, and Jürgen could use them as well should they need to work here, but Ryan didn't think they would be around much." Ben glanced at Camila and grinned. "In Jürgen's case, at least not to work."

Camila frowned. Ben and most of the others didn't know Jürgen was gay, and Jürgen preferred it that way. If they made the wrong conclusion based on Camila's and Jürgen's social activities, Jürgen felt that was a positive. She felt guilty about misleading her friends, but wasn't sure that Jürgen was unjustifiably worried.

Ben showed her the Server Room and the powerful computer that would manage their networking needs. Camila's office was next, at the end of the corridor. "Ryan thought this room would be the quietest." Ben opened the door and urged her forward.

Camila could feel the skin around her mouth stretch into a wide smile. The room wasn't huge, but it was more than adequate for what she would need. It already contained a large, curved desk and an expensive-looking chair. A large computer screen and keyboard dominated one side of the desk, leaving plenty of working space.

Camila gawked at bookshelves and cabinets with more places for storage than she could ever use. A transparency-adjusting window facing south would provide comfortable natural light for most of the day. One corner had a small refrigerator and a coffeemaker.

"It's wonderful. Now that Dani isn't rooming with me, maybe I should live here."

Ben chuckled. "I think there's still room for a bed. Of course, we'll have to move to fancier quarters when EU becomes a billion dollar company."

"I guess this will do until then."

"I have one more room to show you. It's right across the hall."

Ben crossed the corridor and opened another door. "This will be the workroom."

The workroom was twice as big as her office. Her first prototype, rebuilt to replace components damaged in earlier testing, sat on a bench against one wall. A second large bench with chairs filled the center of the room. Tools and equipment similar to what the college had supplied during her research hung on the walls, sat on shelves, or were stored in cabinets. Except these were new and shiny, not worn from long use.

"Can we afford all this?" She walked around the room, touching the large devices and picking up some of the smaller tools.

"Ryan felt we could only beat any competition if we had the best of everything." Ben paused and stood in front of the prototype. "We can build prototypes here from parts fabricated at the machine shop, but we'll have to test elsewhere, of course. Ryan is still looking for something, but we shouldn't need it right away."

Camila nodded. She heard Gabby arriving and took one more look around. "Let's go meet your wife. Then I guess we'd better get to work."

"Jürgen is planning on coming by late this afternoon," Gabby told them when they joined her. "He wants to see the new place, but he also wanted to talk to you about an idea he had."

"What about?" Camila asked.

"He told me to let him tell you." Of course, that made Camila more eager to hear Jürgen's thoughts, but Gabby only smiled.

J ÜRGEN GOT TO THE new facility a little after five that afternoon.

"Gabby said you have an idea," Camila said after quick greetings.

Jürgen smiled. "Take me on a tour first. What's the rush?"

Camila frowned at him and opened her mouth to say something, but hesitated for a second. "Fine. Keep me in suspense." She waved her arm around. "This is the reception area." Her frown changed to a sly smile. "Maybe I can talk Paige into leaving V-World and working for us."

Jürgen shrugged. "That's not my problem anymore. But she's been more tester and less receptionist lately."

The facility wasn't large, and Camila hurried him through a cursory tour. Half an hour later, she sat with Jürgen and Ben in a conference room.

"You've been talking about how difficult it is to come up with a reliable reactor design," Jürgen started. "I think I might be able to help."

"We needed an experienced engineer," Ben said. "For now, Doctor Curtis will work with us."

Jürgen nodded. "That's good, but maybe not enough. For your prototype, you depended on simulations. Gabby's work helped with that."

"Right," Camila said. "As our designs get more complicated, that simulation software is showing its limits. Without support from the school, getting time on a quantum computer is more difficult as well."

"Maybe I can help with that. If we made Gabby's code part of *A World of Your Own*, the sophisticated graphics and advanced code could do better simulations."

"It's just a game," Ben protested.

But Camila looked thoughtful, perhaps thinking about how her first experience with Jürgen's software had been an embarrassingly realistic simulation of creating a company. Jürgen pushed on.

"I programmed my software to use Newtonian physics and a little special relativity, but no quantum mechanics or general relativity. You would have to help me add anything relevant to your work before it would be useful."

"Quantum mechanics certainly," Camila said.

"General relativity would also be important," Ben added. "Remember what Stenhouse said about gravity affecting the field?"

Camila nodded. "Maybe. But what about the new owners?"

Jürgen waved a hand. "It'll be easy to convince them of the favorable publicity this will get them when you're successful." He grinned. "Of course, Energy Unlimited will have exclusive use of the added data and an interface to a quantum computer, at least at first."

44

I T WAS A LITTLE after nine-thirty when Dean Curtis arrived at the Energy Unlimited offices for the ten o'clock meeting. Everyone else was already there, but he had time to get a cup of coffee, a pastry, and brief introductions to the other attendees.

The conference room was austere, but there was a large table and a dozen chairs. Dean took a seat and surveyed the others as they found places. Except for Ryan Terry, they were all so young. *Even he's at least thirty years younger than me.*

He'd already forgotten some names. The Indian girl—Martina?—looked familiar, probably because she had been one of his students at some time. She was a recently graduated mechanical engineer. The girl with Ben Pinto, introduced as his wife, was a software engineer, and one of the boys was a materials engineer. He knew Camila and Ben, but the rest were jumbled together in his mind. He had never been good with names, but he would get them all straight eventually.

Ryan Terry took control of the meeting immediately. He was the first investor in Energy Unlimited and Camila Lopez's sponsor. Ryan stood and talked briefly about the company before turning to their plans.

"You've all seen Camila's concept for a second prototype," Ryan said. "That's way above my head, but Doctor Curtis is here to consult with us on the design." He sat down. "Dean, you have the floor."

Dean nodded and leaned forward with his hands folded on the table. "Some of you may know that I worked in the nuclear power industry for a few years out of college. When fission plants fell out of favor, I went back to school for my PhD and became a teacher, but I've always been interested in nuclear energy. When construction started on the first fusion power plant in California, I consulted on that project, but, as you know, fusion power turned out to be expensive to implement, so only a few facilities were built."

He scanned the faces around him and smiled. "I look forward to working with all of you on this new idea for the generation of nuclear power. The second prototype you are planning will be crucial to your success, exploring the use of solid fuel and the use of magnetism to remove end products before they degrade the reaction. Eventually, we'll have to consider cooling also, but I think we can put that off for now. The initial goal will be a device that can produce more energy than it requires, and do it for an extended time."

P ROFESSOR McGAVOCK HADN'T TALKED to Stenhouse since June and needed to make sure he wasn't forgotten. When Stenhouse came to the phone, McGavock could see he was excited.

"Our version of the prototype is a success," the physicist told him. "In fact, our power yield was higher than Lopez reported in her thesis data."

"That's good news," McGavock said.

"General Power is so happy with the results that they've made me the leader of their project. With a sizeable increase in salary, too."

"You're in charge? Congratulations. I hope you won't forget how much I helped you get support from the government."

Stenhouse frowned. "Well, not completely in charge. They made me chief engineer, but one of their vice presidents is technically in charge. I make all the important decisions." He paused. "We haven't received any funds from the government yet. Senator Correia is working on it."

"He should find it easier now that you have a working prototype."

Stenhouse nodded, and the frown faded. "Yes, I would think so."

"I could talk to him again and push him a little. This is an election year."

"No, I don't think that will be necessary. Giving him a reminder wouldn't hurt, though. I'll call him today."

"It's no problem for me." McGavock tried to keep his voice calm. "After all, I'm the one who originally told him about it."

"I know. For now, I think it would be better if I talk to him. Funds would be allocated to General Power, and you don't even work for them."

McGavock gritted his teeth as Stenhouse ended the conversation and disconnected.

S ENATOR CORREIA HAD TO stop to recall who Arthur Stenhouse was. After all, it had been a couple of months, and he had other problems. His attempt to blame homosexuals for the country's falling population had resonated with some members of his party, but the best available political estimates said that he was still in a tight race for his Senate seat. He couldn't be sure whether the issue had helped or hurt his campaign.

When he remembered who Stenhouse was, he told his secretary to connect the call. Being associated with a solution to the clean energy problem would look good to his constituency. With the election only a month away, Stenhouse's project might be too late to help him much now, but he had to hope he would still be around to take credit in the 2058 election.

"Good afternoon, Doctor," Correia greeted.

"Good afternoon, Senator. I'm calling to get the status of the funding request for my project."

"I'm afraid these things move slowly, Doctor." Correia raised his shoulders in a shrug that was probably not visible to Stenhouse. "I hope to have good news soon."

"Our work to provide safe, clean energy would proceed more quickly if we were assured of funding."

"I understand. The Department of Energy was excited when I told them about your project. I'm a little surprised they haven't pushed for an appropriation yet, but I'm sure they will soon."

"Perhaps you could remind them."

Correia nodded. "Yes, I'll do that. And when I have news, you will be one of the first to know."

After the call, Correia sat at his desk, body back against the chair and eyes staring unseeing at the wall. He could try his contact in the Department of Energy again, but he knew the man didn't like being pestered. A bureaucrat in the department for fifteen years but new to his position, the man didn't have to worry about election campaigns.

If Stenhouse's idea really was a good one, the Department of Energy would support it eventually. Of course, "eventually" could mean years. Still, a press release from his office might have some effect. If he made a favorable comment about the department, it might even goad them into acting without him being resented for pestering them.

45

R YAN HAD INSTRUCTED HIS workstation to watch for news on several subjects and keywords. The computer tagged Senator Correia's press release about Stenhouse with priority high enough to interrupt his perusal of an investment financial report.

The press release urged the Department of Energy to act on the possibility of using the Stenhouse Field to generate power and mentioned that the field's discoverer, Doctor Stenhouse, was working with General Power toward that end. It also told of the senator's urging for action and that, as a result, the Department of Energy was considering a grant to General Power to advance the work.

It was bad enough that General Power was an enormous company worth billions; the government was considering helping them? Ryan didn't blame Senator Correia or the Department of Energy. When he talked to Correia three years before, the senator showed little interest. Ryan had the impression climate hadn't been a big concern, but now, with an election only weeks away, the politician apparently saw it differently.

Energy Unlimited was too small to be considered as seriously as General Power, but it wouldn't hurt to try to get some attention. Over the next hour, he drafted a description of the work they were doing, emphasizing that, although Stenhouse had discovered the field, Camila Lopez had done the research showing its potential for power generation. Then he asked Dani to get contact information for the senator, the Department of Energy, and the media, and send out the message as soon as possible.

E VERYONE MET IN THE conference room for pizza and discussion the following evening. "I'm not sure what more we can do other than make Energy Unlimited more visible," Ryan told the group. "My release suggested that the government could fund us, perhaps even instead of General Power, but I don't expect that will be taken seriously."

"Do we need the money?" Gabby asked. "We have my brother's investment now."

"Not immediately," Ryan said.

Dean Curtis jumped in. "But you will soon. This group suffices to build prototypes. But you're going to need more people, especially engineers, to go past that. To sell a product, you'll need manufacturing capability."

Ryan nodded. "You're right, of course. The real competition with General Power is in producing the first product, not just prototypes."

"We're not getting anyplace looking for more investors yet," Mike said. "Having the first successful prototype wasn't the boost we hoped it would be."

"We could push harder for government funding," Dani said.

"And we'll do that," Ryan replied. "But with General Power in the picture, it's probably a waste of time." Martina was frowning and staring at her hands. "Martina, you look like you have something to say."

She looked up with a start. "I don't know. Maybe."

Ryan held out an open hand. "Go ahead. We're all friends here."

Martina hesitated and bit her lip. "OK. I think you know I'm here on an educational visa. I'm still an Indian citizen. The government of India has paid for most of my expenses while I get my degrees. In return, I occasionally report on the public aspects of what we're doing." She placed a heavy emphasis "public."

She paused and looked around the table until the murmurs faded. "That's not required. Mostly, they just want me to come back with an engineering education and work on India's infrastructure projects. But it gives me an idea. India might be willing to help fund our work."

"What would they want in return?" Ben asked.

Martina shrugged. "I don't know. I guess that would be something you would have to negotiate."

"It's a possibility," Ryan said. "Do you have any contacts in the Indian government?"

"Only low-level contacts in the Ministry of New and Renewable Energy. They could probably talk to higher officials."

"That should be good enough." Ryan smiled. "Pass the word that we would be open to discussing their needs and how we might address them in the long term." He stopped for a second. "Assuming there are no objections from any of you."

"It doesn't hurt to talk," Ben said, and several others nodded.

"YOUR BIGGEST ISSUE IS cooling," Dean told Camila and Ben. "Relatively cheap materials like lead and tin have low melting points. Keeping a solid fuel core from melting will be the limiting factor on how much energy you can produce."

He cautioned himself not to sound too pedantic. That had always been a problem in his teaching career. His students saw him as gruff and intimidating, not what he wanted. He enjoyed teaching and being around young people eager to learn, but his manner had been an obstacle.

"What's the melting point of lead?" Camila asked.

"Around 620 degrees Fahrenheit," Dean answered. "Tin is even lower."

Ben took out his reader and entered a command. Then he frowned at the answer. "That won't work. No matter how much cooling we provide, the energy release won't just melt the fuel. It will vaporize it."

Dean nodded. "Exactly. You must control the reaction by controlling the available fuel."

"Martina said that existing fission power generation uses rods of fuel," Camila said.

"With expensive cooling systems." Dean smiled. "The fuel is generally encased in ceramic and metal. Power is generated from neutron emission, not the binding energy directly."

"The fuel would have to be in contact with the Stenhouse Field," Ben said.

"Any fuel will be instantly vaporized," Dean agreed. "So don't try to prevent it. Use it. Have a conventional first stage that vaporizes the fuel and then injects it into the Stenhouse reactor to produce usable energy. A tiny amount of that could be diverted back to power the first stage and the magnetic field that will remove the reaction byproducts."

"If the first stage is powered by the second stage, you would need to jumpstart the reaction," Ben said.

Dean bowed his head briefly. "A storage battery could supply enough energy to create the field, but not enough to vaporize solid fuel. The second stage wouldn't care what is used for fuel, as long as it's heavier than iron. Injecting a little krypton or xenon will easily get enough power to start the first stage once the field is established."

"This all sounds complicated," Camila said.

Dean smiled at the two young physicists. "It usually is. That's where I earn my pay. But we can develop the components separately. For example, we could test a recovery stage to collect the residue with a low energy generator—maybe even your first prototype." He raised his hand and pointed upward with a thoughtful expression. "The recovery stage might even be patentable on its own."

"Ben and I have been working with Jürgen to train his software about advanced physics," Camila said. "If we can use that to get better simulations, it would help."

Dean didn't express his doubts. Simulation software could do a lot, but they had to rent time on a quantum computer, limiting what they could do.

He didn't want to disappoint the two fledgling physicists. They had made incredible progress, but now it was time for engineers to take over. They would have to accept that, and he hoped they would realize it on their own if they hadn't already.

P RISCILLA CORREIA WOKE SLOWLY. Her husband had let her sleep, and she appreciated the consideration. The state televised continuous coverage of elections on a public channel, and Leo would be watching in his office. He could be obsessive about the returns, but Priscilla was more patient, knowing it would be hours before enough votes were counted to allow realistic predictions.

She turned on the bedroom television and selected a station that offered general morning news. A reporter was interviewing some scientist about global warming, with the scientist apparently trying to make a case that it was still a problem.

"If your claims are correct, why is the government saying the problem has been resolved?" the reporter asked. He leaned toward the scientist with a skeptical sneer.

The scientist's reply was almost a growl. "Because no one respects science anymore. We scientists deal with evidence, but politicians and too much of the general public prefer wishful thinking."

The reporter scowled and thanked the scientist for his time. The broadcast then shifted quickly back to the studio and the news anchor.

"Thanks for that report, Dan." The anchor stared out of the screen. "We now have an update from Marcus Campbell on the demonstration at the Jackson Street polling place in Brooklyn."

The view changed to a reporter standing to the side with a crowd of shouting people behind him. The reporter explained that gay demonstrators had assembled to protest, and the confrontation had begun when counter-protesters showed up.

Priscilla could hear angry exchanges as the camera moved closer to the protestors, close enough to see that some of them wore pins, hats, or shirts expressing support for or opposition to her husband's election. It was not surprising, but she was still shocked to see that the homosexuals opposed him while their opponents favored Leo.

She shuddered and turned off the television. Then she pulled the covers up to her chin and tried to bury her head in her pillow.

46

S AHIL NANDA HAD A bachelor's degree in chemistry and kept up on technological advances, but he'd gotten his master's in business and gone to work as a manager for the fledgling Basu Power upon graduation. Since then, he had risen rapidly, becoming the Chief Technology Officer when only fifty-five years old, two years before. The company had risen with him, outperforming other Indian corporations with aggressive pursuit of innovative technology.

He learned of the Stenhouse Field's potential for power generation through reports on United States energy research and General Power's plans. Before that, the Stenhouse Field was an interesting discovery with significance in astrophysics, but little else. General Power ignored overtures about investing in their efforts. Facts were scarce, but rumors circulated about support from the U. S. government and the German company *Saubere Energieerzeugung*.

India was dedicated to developing its economy without destroying the environment. There were thousands of companies working on solutions to India's energy needs, with little help from the west. Most of those companies were older than Basu Power, and many were bigger, but there was no sign they had gotten anywhere with General Power either.

It was possible that one of the hundreds of engineers working for him knew something useful. "Search for internal documents on Stenhouse," he told his workstation.

There wasn't much, but a report from the Ministry of New and Renewable Energy caught his attention. An Indian woman studying mechanical engineering in the United States had reported being hired by Energy Unlimited, another company working on using the Stenhouse Field to generate power. Sahil couldn't find much on Energy Unlimited, probably because the company had existed for only a few months. He did find a press release on the company's founding saying that Camila Lopez, a physicist, had performed research on exploiting the field and was one of Energy Unlimited's founders.

He knew little more than what Martina Patel had told the Ministry, hampered by the data restrictions imposed by the United States. With financial resources the smaller company couldn't match, General Power would be favored to win in the race to develop products that could dominate the market. But if Energy Unlimited had the money as well as the developer of the technology, they could beat the odds.

Energy Unlimited could be an opportunity for India in general, and Basu Power in particular, to gain access to an important energy source.

"YOU WOULD THINK THEY could handle this better now," Leo Correia told his wife. "The election was four days ago, and they've declared me the winner only this morning."

"It was very close," Priscilla said. "And some people still use mail-in ballots."

Correia shook his head. "It was close because the gays raised so much antagonism toward me."

Priscilla put a hand on her hip and frowned. "Well, what did you expect, Leo? You used them to drum up support, and they were hounded because some idiots listened to you."

"I only told the truth. They don't contribute to society and should be made to."

Her eyebrows went up. "You're actually starting to believe that crap?"

"You're supposed to support me, not whine about them." He could hear pleading in his voice.

"Them? Who's them? Other human beings with different lifestyles than you?"

Correia sighed. "We used to think alike. When did you turn into a liberal?"

Priscilla turned away. "I'm not the one who changed, Leo."

RYAN CALLED MARTINA FROM his office. As he expected, she was at the machine shop, working out the best way to fabricate parts for the second prototype.

"It looks like your contact in India came through," he told her. "Basu Power sent a message asking if we would talk to a representative."

"I've heard of them. They're a new company, but I don't know much about them."

"I guess we'll find out more soon. Someone will visit next week. I'll need you at the office when they're here."

"Not a problem."

Amrita Savant had not been to the United States before. She knew intellectually that its population density was much lower than India's, but Albany was the capital of one of the most populous states. New York City accounted for much of the state's population, but her brief trip from Albany International Airport to the Energy Unlimited headquarters in Schenectady still surprised her.

She was early for her meeting at Energy Unlimited, so she had told her taxi to use a scenic route. Her reader told her that the automated vehicle was taking a less-traveled route closer to the Mohawk River. Perhaps the main road would have been different, but she saw more wooded landscapes than houses, and the dwellings she saw were single-family homes surrounded by an astonishing amount of land. It was late November, and the trees were just starting to take on fall colors.

Schenectady seemed more familiar to her. Warehouses and other commercial establishments dominated the area where the taxi took her. The building it stopped at was typical, although looking better maintained than some of its neighbors. No sign confirmed it as her destination, but it was the right address.

The taxi had notified Energy Unlimited when it was a few blocks away. As Amrita walked across the small parking lot, a man and a young woman obviously of Indian heritage came out to meet her.

The man held out his hand. "I'm Ryan Terry, and this is Martina Patel, one of our engineers. Welcome to America."

Amrita shook his hand, and they went inside. A woman Amrita recognized as Camila Lopez joined them, and Ryan led them to a conference room where a coffee service was already set up.

For the next hour, Amrita listened to them talk about their plans. Camila spoke in general terms about how they were harnessing the Stenhouse Field to generate energy, Martina described her work at the machine shop they had purchased, and Ryan talked about the company's finances.

"We have sufficient funds from other investors for the near term," Ryan finished. "Eventually, we will need more to actually develop a product."

Amrita frowned. "My company had hoped we would be the only energy company to invest in Energy Unlimited. Much of our interest is because of that."

Ryan smiled. "Our largest investor is a software game designer. Right now, I'm the only other investor." He paused. "Of money, at least. Camila, Martina, and several others have heavily invested their time and intellect."

"I would like to meet the others as well."

"Of course. I suggest we get acquainted over dinner tonight. Do you have a preference for the type of cuisine?"

"I'm quite cosmopolitan. But it has been a long day for me. Perhaps I could go to my hotel and rest for a few hours?"

"That's not a problem." Ryan stood. "Let me take you over there. It's not far from here. Then I'll pick you up around eight?"

Amrita glanced at her watch. Ryan's suggestion would give her a welcome five hours. "That will be fine. I look forward to meeting everyone then."

As she followed Ryan to his car, she considered her options for pressing her goals. Ryan was probably about twenty years older than her, and she could tell he found her attractive. He wasn't wearing a wedding ring either. That was promising. From what she had been told before coming, he was the oldest of the major players in Energy Unlimited, and possibly vulnerable to the kind of persuasion she knew she could supply.

The briefing they had just given her had been encouraging. Camila Lopez was the technical linchpin of their operation, but she would listen to Ryan Terry. If Energy Unlimited could really do what they hoped, using money from Basu Power, the benefit to her company and therefore herself would be immense.

47

D ESPITE AMRITA'S CLAIM TO be open on dining choices, Ryan put some thought into a selection. A steakhouse was probably a poor choice and, although he knew several Indian restaurants he liked, he feared they might not measure up to the real thing. Ryan finally decided on Italian and texted the others to meet him there.

Ryan used his own car to pick up Amrita. Like the taxis, it was automated, but more comfortable, with a sound system programmed to play music Ryan chose. He chose a selection of soft classical music, playing at a volume that wouldn't inhibit conversation. Amrita had already seen their offices, but the machine shop and Camila's test facility were nearby, and he pointed them out as they passed them.

At the restaurant, he went around the car to help Amrita. She smiled and took his arm as they walked across the parking lot. She pressed closer to him than was necessary, but he wasn't surprised.

The dinner went well, more social than business. With everyone there, twelve people including Mary Welch, serious discussion would have been impossible in the busy restaurant. It was more important to introduce everyone and convince Amrita of their purpose. Amrita seemed to enjoy the food and the people.

When he took her back to her hotel after dinner, she again took his arm as he escorted her into the hotel lobby. They stopped there.

"I can pick you up tomorrow," he told her. "What time would you like to begin our talks?"

She stared at him for several seconds, and he expected her to invite him to her room. That would not be wise, and he was thinking about how to put her off gracefully, but she released his arm.

"I'm sure you're very busy. I'll take a taxi. About eight o'clock?"

Ryan smiled. "That will be fine. Have a good night."

She nodded and returned the smile. Then she turned and walked toward the elevators.

W HEN RYAN GOT BACK to his home, Mike and Dani had already let themselves in and were waiting for him. Unsurprised, he ushered them into seats in his living room.

"Dani wasn't sure you would be home tonight," Mike said.

Ryan chuckled. "I know why you think that, but here I am."

"She wants to manipulate you into favoring her during negotiations," Dani said.

"I know. So far, she's being subtle, and she might not go further than that."

Dani stared at him. "And if she suggests more?"

Ryan shrugged. "She might think she can influence me that way, but it won't work." He looked at his son. "You didn't think she could get to me that way, did you?"

"Not me," Mike said. "We Terry men are resistant to that kind of thing."

"Really?" Ryan noticed the sly smile on Dani's face. Dani caught his gaze and winked.

Mike must have noticed something, because he turned red and looked at his wife.

"We've both had experience with women who thought they could exploit us for money," Ryan said.

Mike nodded. "I remember you stepping in once."

"Yes, I have a sense for the kind of woman who wants to take advantage of us, and Amrita Savant fits the bill." He looked at Dani again. "I realized early Dani didn't."

Mike stared, speechless, at his father, but Dani only laughed and kissed Mike on the cheek.

"Anyway, don't worry about me, Dani. I think Basu Power wants us as much as we want them."

R YAN HAD ASSURED DANI Amrita would not take advantage of him, but he was less confident negotiations would result in an agreement. The other four board members agreed they would not reveal significant technological details. Basu Power would not like that.

They also agreed he would be the chief negotiator, and he asked Ben to assist him. Ben could probably answer technical questions they were willing to answer, and Camila would be nearby if needed.

Amrita arrived a few minutes early, wearing a silk dress that clung appealingly to her generous figure and strategically exposed just enough of her breasts. Ryan kept his expression neutral, but friendly, as he greeted her and escorted her to the conference room where Ben joined them.

Predictably, technology transfer was at the top of the list of issues, and by the end of the first day, little agreement was reached. Except for a brief lunch break, they argued the same points all day, and by late afternoon, everyone was exhausted. Camila joined them at the end and Jürgen and Gabby came from V-World to learn how the talks had gone.

Ryan expected Amrita to make some suggestion about the evening, but he was amused to see that Jürgen distracted her. Admittedly, Jürgen was her age, but Ryan felt a little disappointed. While they said goodbyes, though, Jürgen asked Camila out to dinner, and Amrita frowned, probably in frustration. Their relationship had not been obvious at the restaurant the night before.

"Again, tomorrow at eight?" Ryan asked.

Amrita nodded and used her phone to call for a taxi.

B ACK AT HER HOTEL room, Amrita called Sahil Nanda, Basu Power's Chief Technology Officer. He listened without interrupting while she told him of the day's progress and what she hoped to accomplish on the next day. She didn't mention her lack of success in influencing Ryan. Ryan had been attracted to her, but she had allowed herself to be distracted by Jürgen. Her chances with Ryan were gone, and Jürgen was in a relationship with Camila Lopez and had shown no interest in her. At least Sahil didn't bring that up, saving her from the embarrassment.

"You've done well so far," Sahil said when she finished. "Their offer to give us the rights for Pakistan as well as India is a welcome surprise that will give us leverage with the government."

Relations between India and Pakistan alternated between distant and hostile. If Energy Unlimited was successful, controlling how Pakistan used its technology could be extremely useful.

"Licensing fees for the technology are undetermined, of course, since we don't know what we'll be licensing until they have a product, and they won't give me details."

Sahil nodded. "Realistically, their desire to have secrets is understandable. Until they have a patentable design, their competitor could steal their work and market products before they can. Part ownership in Energy Unlimited will be worthless if that happens."

"We have broached the subject of investment without going into details. Perhaps tomorrow."

On the phone screen, Sahil nodded. "I have discussed your mission with the board. We agree that this new technology could be a game changer for our company. We are prepared to aggressively pursue getting a stake in Energy Unlimited."

"General Power has an advantage over them and is likely to beat them to marketing a product."

"The board is of the opinion that General Power's biggest advantage is money. We are prepared to remove that advantage." Sahil smiled and told her how much Basu Power was willing to invest.

48

A MRITA ARRIVED PROMPTLY AT eight, and Ryan was amused to see she had chosen more modest attire. Before she arrived, Ben confided that Gabby had wondered about Ryan's reaction to Amrita. They joked about it, with both men suggesting that the other had been interested, and Ryan was glad he had kept the encounter professional.

They spent the morning resolving the remaining issues about licensing, made complicated by not having any actual product to license yet, and Ryan's insistence that EU would not release technological details. Amrita's arguments seemed more subdued than the day before, though, and after a lunch ordered from the nearby coffee shop, they began talking about money.

Ryan had paid close attention to Amrita during their discussions. In college, he had had some success playing poker, and he used those skills to read Amrita. He had expected more pushback on the licensing issues, and saw the Indian woman's hesitance as a sign of how badly Basu Power wanted to make a deal. An investigation of the company before she arrived suggested that Basu management was willing to take risks for potentially large returns.

He made his opening proposal based on that. "So, under the licensing terms we've agreed to, Energy Unlimited is offering twenty percent of the company for six hundred million dollars."

"We would want representation on the board," Amrita said.

"We can give you a seat."

"That's a lot of money for what is a very uncertain investment. Four hundred million for thirty percent and two seats on the board."

Ryan had difficulty keeping his expression neutral. Amrita's counter confirmed his belief that Basu was eager to make a deal. There were three quantities being negotiated, and he had to balance them. Two seats on EU's board were acceptable, although one would be better. Either way, the existing five members would still have control. Thirty percent of the company was within the amount they had set aside for investors, but less would leave the possibility of additional investment.

They haggled over the details for the next hour. "Five hundred million for twenty-five percent of the company and two seats on the Board," Ryan finally said. "Subject to a vote of the full Board, of course."

"How soon can you meet to vote on it?" Amrita asked.

Ryan looked at Ben. "Can you speak for your wife?"

"I would rather not. But she'll be here in an hour or so anyway."

Ryan nodded. "Jürgen is picking Camila up soon and everyone else is already here, so I can present it tonight. If there are no objections, we should have a decision in the morning."

C AMILA WAS TRYING TO concentrate on the screen of calculations in front of her, but it was late in the day and her focus was fading. She heard exuberant shouts from outside her office, saved her work, and walked out into the hallway. Ben and Ryan were near the exit, giving each other high-fives and talking excitedly.

They turned as she approached, and Ryan punched the air. "Five hundred million for twenty-five percent."

Camila was speechless. Ryan had been encouraging before the day's negotiations, but she hadn't expected this. If Jürgen's fifty million was a huge step forward, ten times that was a leap to the heights. *How can we use that much money?*

"I told Amrita she would have a tentative agreement on our part in the morning, after the full board has heard the offer," Ryan said. "After that, it will be up to the lawyers to draw up the contract."

Camila's mind churned as elation and astonishment competed with fears that she would fail her friends because she wasn't up to the challenge. They went into the conference room, where Ryan and Ben discussed the deal enthusiastically, but she was too bewildered to participate very much. The two men had calmed down when Gabby and Jürgen showed up, and it all began again.

Jürgen seemed to understand her feelings and took her aside. "This is fantastic news. I guess I got a great deal with my investment." He smiled. "It's a lot to take in, though. Let's get away from all this and go out for a nice, quiet dinner."

Camila looked at the others. They might be in a mood to celebrate, but Jürgen's suggestion had more appeal. "That sounds good," she told Jürgen. She turned to the others. "Let's plan for a big celebration tomorrow night. We can get everyone else together then."

Everyone seemed to agree, and she left with Jürgen.

D EAN CURTIS LEANED BACK from the computer terminal. "This looks good," he told Martina. "I think it's ready to give to the machine shop."

Martina smiled. "I couldn't have done it without your help."

"Separating the final products from the elements still fissioning was tricky." He stood. "You would have figured it out eventually."

"Maybe. Do you want me to send this to the shop?"

"I haven't seen your machine shop yet. Why don't you copy everything onto a memory stick, and we'll deliver it in person?" He looked out a window and saw patches of bright blue sky between the adjoining buildings. "It's a beautiful day. Let's walk over there."

Martina looked doubtful. "It's more than half a mile from here. Are you sure?"

"The distance from the parking lot to my office on campus was almost that far."

His tone was gruff, intending mock severity, but Martina frowned, possibly misinterpreting his mood. He smiled, and her expression cleared.

"All right," she said. "Give me a couple minutes to copy everything."

M ARTINA KEPT UP WITH Curtis, despite her shorter legs, but it was an effort. The route between the offices and Welch Precision Machining took them through an industrial area, but the rain from the previous day had passed, evolving into a sunny December day. The air had cleared enough to keep their masks in the pockets of their light jackets.

"When I was your age, there would be snow on the ground in December." Dean looked up at the sky and smiled. "It was colder, though. This is much better for a little stroll."

Mary Welch greeted them cheerfully when they reached the machine shop. Martina introduced Dean, and Mary gave him a quick tour of the operation. While Dean paid attention to the facility's equipment, Martina watched Mary. The older woman's demeanor had changed in recent months from tired and defeated to energetic and positive.

"How's business?" she asked when they were back in the relative quiet of Mary's office.

Mary smiled. "Better, and still improving. The new marketing plan has brought half a dozen customers back and added two new ones. Pretty soon, I'll have to turn work away."

"I'll tell Dani. She'll be happy to hear it."

"So Ryan's son finally married her. I hope he knows how lucky he is."

Martina grinned. "I'm sure she's told him."

Mary smiled and nodded. "Well, let's look at what you've brought me."

Martina handed her the memory stick with the design for the recovery stage casing, and Mary inserted it into the computer on her desk. A couple of commands brought the design up on the screen.

Mary scratched her head. "Complicated. And you want it made of titanium?"

"Reaction products will be very hot," Dean said. "Chiefly vaporized iron."

"How hot?"

"About 2800 degrees Fahrenheit."

Mary frowned. "Minimum? Could it be hotter? Titanium's melting point isn't much higher than that."

Dean smiled, a look Martina had seen when the former teacher approved of a student's response. "About 3000 degrees. But the volume of iron will be small, and the reaction controlled carefully so that the temperature doesn't get too high."

"A tungsten alloy would have a higher melting point."

Tungsten would fission in a Stenhouse Field. They were considering tungsten alloys for future use where components wouldn't be exposed to the field.

"Not for the prototype," Martina said. "There are other issues we need to study."

"OK, if you say so." Mary shrugged. "Anyway, I can do this, but I'll need to build special tooling to get some of these elements within the tolerances. It'll take a couple weeks."

"We expected that," Martina said. "This is the most complex part. I'll have a few smaller, simpler components in a few days."

T HE ELECTRONICS FOR THE first test had been crude, wire-wrapped boards built by Camila with some help from Ben and Gabby, using programmable components. As their devices became more complicated, they would need more experienced help and custom controls. Shortly after the new year, Energy Unlimited added two electrical engineers to their payroll.

Marlee Dubois was an electrical engineer and circuit designer, a cheerful woman in her sixties. Her decades of experience in designing complex electronics for several prominent companies and her desire to try something new made her an ideal candidate.

Emma Martin, still in her twenties, was less experienced, but had impressed Camila, Ben, and Ryan with her energy.

Their first task was designing the electronics for the recovery stage. After a few days to get familiar with the original prototype and the requirements for the new component, they set to work. Although the energy output would only be measured initially, they would eventually augment the recovery stage to convert the energy of the plasma into usable power.

W ITH THE TWO ELECTRICAL engineers working on control circuitry for the prototype changes, Camila and Ben turned their attention to the first stage, where the fuel would be vaporized. Some of the same issues in managing the elevated temperatures in the reaction chamber applied in the first stage, a furnace hot enough to turn fuel into a gas, as well.

Camila stared at the simulation results of the first stage design displayed on her workstation. "This looks good." It had taken two months, but they had finished adding basic quantum mechanics and general relativity to *A World of Your Own*. They could link the program to take advantage of time rented on a quantum computer. The first stage's simulated operation was displayed in detailed graphics.

"We'll have to run this by Dean," Ben told Camila. "I don't think he trusts the software."

It was early March, and the test of the recovery stage was only days away. Camila nodded. "We shouldn't take him away from the test. We'll show him after that."

"This kind of mechanical engineering isn't our strong point, though. We should start leaving these tasks to people like Dean and Martina."

Camila opened her mouth to speak, but then closed it and thought about Ben's statement. "You're right, of course. I don't want to be useless, though."

Ben smiled. "We're physicists. We should concentrate on the test results and getting a better understanding of the Stenhouse Field. I strongly believe that we'll need that." He stopped and put a hand on Camila's shoulder. "What we're doing now is great, but other than propulsion, it doesn't get us any further in building spaceships."

"Other than propulsion? Are you bringing up warp drive?"

Ben shrugged. "I still don't think a Stenhouse Field can be used to go faster than light, but it could be harnessed in other ways than fissioning nonradioactive materials."

"So I think you're suggesting you and I should be the research division of Energy Unlimited."

"Doing what we're trained to do."

49

The second prototype test boosted company morale almost as much as the agreement with Basu Power.

"I expected to still find signs of deterioration from byproducts like bromine," Dean told the group at the test debriefing. "The currents created by drawing the iron out of the reaction chamber caused nonmagnetic products to encounter the field more, resulting in better completion."

"How well did it do on removing the end products?" Ryan asked.

"Pretty well, I think." Dean shrugged. "For this small output, there isn't much. I can only see a slight sheen at the deposit area, and none where the hot gases enter, but I wouldn't expect to. We'll know more when we have the parts analyzed. Let's rush that. I would like to run the test again using a stronger Stenhouse Field."

"That would help me get better data on the reaction chamber material," Rafael added.

Dean nodded. "That will be critical at higher power levels. We need better data on shielding requirements, too. Neutron emission was higher for this test."

Ryan leaned forward and looked at Dean. "Do we have enough for a patent?"

Dean gave the question a moment's thought. "I'm not an expert, but I have a little experience there. We might. Certainly enough to start preparing a patent application."

"Ben and I can work on that," Camila said. "I assume building the prototype for the first stage is the next step."

"I recommend the further testing I mentioned of what we already have," Dean answered. "We can work on the first stage at the same time. We may need to redesign the reaction chamber as well."

"Ben and I have run the simulations on the first stage design. We think it's complete."

Dean frowned but didn't reply.

"Do we need more people yet?" Ryan asked.

Dean shook his head. "Not yet."

"Okay. I don't think we should hold back right now, though. Our best hope is to move faster than General Power and get a few patents on what we develop."

Dean nodded. "Ultimately, a few good patents could be the difference."

J ÜRGEN STILL MANAGED V-WORLD, as stipulated in his agreement to sell the company. As a result, he wasn't at the test or the debriefing afterwards, but he had dinner with Camila that night, and she told him about it.

"Sounds like you're busy." Jürgen smiled when she finished. "I'm glad you could make the time for me."

Camila returned the smile. "Sometimes even I need a break. And patent applications are not exactly what I had in mind when I decided to study physics."

"I sympathize. I spend a lot less time coding now than I used to. You wouldn't think the sale would make a difference, but it has."

"We've used your software for the first stage design. Ben and I were impressed, but Dean has doubts, I think."

"The first stage vaporizes the fuel, right?"

Camila nodded, and Jürgen continued. "Dean will feel more confident about the simulations when you can compare them to the test results."

Jürgen took a bite of his food and Camila also returned to her meal, and their conversation paused. Jürgen broke the silence a few minutes later. "Have you had any luck finding a roommate?"

"I've been too busy to try." She shrugged. "That's another reason—besides your company—why I need to make time for other things than work."

"I guess it's lonely now that Dani is gone."

Camila nodded. "I miss her. There's no one to talk to at night."

Jürgen didn't answer for what, to him, seemed like a long time. He knew what he wanted to say, but the possibility that it would hurt their comfortable relationship held him back.

"I've had an idea for a while," he said, trying hard to speak slowly and deliberately. "I have this big house for just me, and I wouldn't mind having someone to talk to either."

Camila's eyes went wide, but he couldn't tell whether it was from shock or surprise.

"Are you suggesting I move in with you?"

"I have four bedrooms—well, one is an office—and you know you'd be safe with me."

Camila shook her head and didn't speak. Her eyes closed, and he hoped it was because she was thinking. After a long few seconds, the corners of her mouth curved upwards in what might have been a smile.

"But would you be safe with me?" And now she was definitely grinning.

This was the moment he had most feared, but he plunged ahead. "I think I would be as safe from you as you would be from me. And for the same reason."

This time, he was sure her expression showed shock. "You think I'm gay too?"

He spread his hands and inclined his head slightly, but didn't speak. She pressed her lips together and looked down, not at him.

"Is that what's wrong with me? Why I can't be close to a man?"

"There's nothing wrong with you, and I think you and I are close. That's not what you meant, of course." He reached across the table and took her hands.

Camila sighed. "I've got to think about this."

The rest of the dinner was awkward, but still friendly at least.

A FTER JÜRGEN DROPPED HER off at her apartment, Camila should have gone to bed, but she knew she wouldn't sleep. She sat in her living room, staring at a picture on the far wall without really seeing it. Jürgen had to be right.

Dani tried to tell her once, but she had ignored Dani and so many other signs. Jürgen saw them, but was that because he too was gay? Or did others know?

Her relationship with Jürgen was based in part on his desire not to reveal his sexual preferences. There was a genuine bond between them, but his fears were real. Greg had accused her of being gay, probably in anger rather than any actual insight, but others would notice her lack of interest in men. For the first time, she understood what Jürgen was afraid of.

It was all too much to deal with. She needed to talk to someone, but who could she talk to? The answer was obvious, but Jürgen was suggesting more than just talking with his invitation. The implication was that she needed cover, too. Moving in with him would lead to unwelcome interpretations, and she would have some explaining to do with her family. But perhaps that was better than the alternative.

It had been a long day, and she was tired. If she could get some sleep, she might consider the situation with a clearer head.

50

At Camila's request, Jürgen came to the Energy Unlimited office to take her out to lunch. It was a warm day in mid-March, and they decided to walk to a small restaurant a block away.

"I'm thinking seriously about your offer," Camila said as they walked. "I have a question."

Jürgen smiled. "At least you're thinking about it. And I'm a terrible cook. I was hoping you knew how."

Camila grinned and shook her head. "My mom made sure I could manage in the kitchen, but that's not my question." Her smile faded.

Jürgen looked at her. "Oh, this is a serious question."

"I would think that relations with other gay men would be an issue. How would that work?"

Jürgen frowned and didn't answer for several seconds. Then he spoke slowly and hesitantly. "I'm not currently in a relationship and don't expect to be in one. I did have a lover several years ago, but that's over."

Camila could see that the question troubled Jürgen and debated dropping the subject, but Jürgen shook his head and continued speaking.

"I'm only telling you about it because I'm not sure you understand the dangers of being gay in the current climate. The government pays lip service to gay rights for the most part, but there are too many hostile people out there."

Camila glanced down the street. The restaurant was only a couple of doors away and was probably crowded with customers, but the sidewalk was clear. She stopped Jürgen and gently pushed him against a wall. "What happened?"

Jürgen sighed. "He went out one night on his own and stopped at a gay bar in Saratoga. The police raided the place and, according to the official report, he resisted arrest. A judge sentenced him to a year in prison, and someone killed him in the first month."

"Because he was gay?"

Jürgen shrugged. "Who knows? But he wouldn't have been there if he wasn't gay."

Other than Greg's angry comment, she had never experienced a problem. But she hadn't admitted to herself that she was gay, so perhaps that wasn't surprising. Jürgen didn't mention it, but people like Senator Correia fed existing bigotry.

"I appreciate your offer. I think I will take you up on it."

Jürgen nodded and smiled. "Good. I think that's the right decision."

CAMILA WOULD HAVE TO tell her parents and didn't expect that would be a pleasant conversation. She could call them, but that would be the coward's way. It had to be done in person. They wanted her to come home for Easter, less than a month away, so she delayed moving until after the visit.

She flew to Texas on Thursday and spent the next two days waiting for the right moment. On Saturday night, Diego went out, leaving her alone with her parents.

It's now or never. They were sitting in the living room, and Camila leaned forward toward her parents. "I would like to talk to you about something."

Nina always knew when something was bothering her, and this time was no different. "You look very serious, darling. Is something wrong?"

Camila shook her head. "Do you remember meeting Gabby Varella's brother at her wedding? Jürgen?"

Nina smiled. "A very handsome young man. I remember."

Tomás scowled at her, but he also put an arm around her and hugged her. "I also remember. Perhaps not as well as your mother."

Camila wanted to return her mother's smile, but was too frightened about how they would react. "Jürgen has asked me to move in with him. I said yes."

Camila hoped for congratulations but expected reproach. But both reacted with puzzled looks.

"Why?" Nina asked.

Camila was too surprised to answer immediately, but Tomás and Nina stared at her, and she could see only curiosity on their faces.

Nina broke the silence. "You've become friends? Friends that close?"

Nina's use of the word "friends" would have been a euphemism for some people, but not her mother. Her parents assumed that she and Jürgen were not lovers. Now, she was sure, she was the one who looked puzzled.

Nina reached over and took her hand. "We know it's not about sex. To put it delicately, he's not your type, is he?"

"You knew?"

"That you're a lesbian?" Tomás said. "Since you were a teenager."

Nina tightened her grip on Camila's hand. "We've waited for years for you to tell us."

Camila sighed. "I've only recently admitted it to myself." She grimaced. "Jürgen told me. He confronted me about it when he asked me to move in with him."

"He knows you're gay, but he still wants you to live with him?"

Camila nodded. "We've become best friends. And we have the company in common."

"And he's gay too," Nina said.

Camila nodded and sat back in her seat as relief flooded through her.

H ER TIME WITH HER family had rejuvenated Camila, but back in New York, the feeling didn't last long. While Dean prepared for another prototype test at a higher power level, she worked with Ben on a patent application for the recovery stage. Ryan had arranged for a consultation with a patent attorney, and the work went smoothly, but Camila didn't enjoy it.

They had the analysis of the results from the last prototype test. The end products collected by the recovery stage were ninety-four percent iron and iron hydrides, with lesser amounts of nickel and cobalt. Trace amounts of other elements were almost undetectable. That was encouraging efficiency for the relatively crude prototype.

Although the end products were safe, the process had created enough neutron radiation to be a concern. It wasn't a problem for the low-output prototype, but practical generators would require heavy shielding. Camila would have preferred to study that issue, but the patent application was more urgent.

After a patent search, which was almost routine for such an innovative application, Energy Unlimited submitted the initial application and received a Patent Pending designation in June. There would be more paperwork, and a final patent was still more than a year away.

Ben could see her frustration, probably because he felt it, too. "Just keep telling yourself that someday this will allow you to work on your dream," he told her frequently. "We're all behind you."

ARTHUR STENHOUSE STUDIED THE results of the last General Power prototype tests and frowned. Zachary Oakes, the project head, wanted to meet with him in ten minutes to discuss progress, and Stenhouse wasn't happy about what he would have to say.

The prototype reactor was working as expected, generating more energy than was put in. Known techniques could easily convert the hot plasma of iron and hydrogen nuclei to usable power. The problem was with the cooling iron that remained. The quantities were small, but enough to clog components if not eliminated.

Stenhouse had worked with engineers on his team to design a solution, but the previous day, he had been told that they couldn't use the design because it would infringe on a pending patent that Energy Unlimited had filed. He wasn't directly responsible for solving the problem, but as the head engineer, he would have to explain the issue to Oakes.

The display of test results offered no solutions, and he hadn't expected it to. All he could do was grit his teeth and face Oakes. He had delayed the confrontation as much as he could, but he was out of time.

The General Power vice president's office was in an adjoining building, and Stenhouse arrived out of breath, but the receptionist told him to wait, and he had several minutes to recover. Oakes looked up at him with an impatient frown when Stenhouse finally stood before his desk.

"I understand the project is falling behind schedule."

Stenhouse nodded. "Part of our prototype must be redesigned because our first design may infringe on an Energy Unlimited patent. That has set us back a couple of weeks."

"Why didn't you know about this before?"

"Until recently, everything they've done has been public knowledge. Once Camila Lopez graduated and formed the company, they've been less open. Now that they've submitted the patent application, we know what they're doing, but we can't use it."

"Can we challenge the patent?"

Stenhouse shrugged. "Maybe. I don't know. That would be a question for a lawyer."

"And we will consult one, but that will take time too." Oakes seemed to settle in his seat as if he were easing from a tensed posture. "We can't let this happen again. You didn't think this startup would be any competition."

"That was before they miraculously came up with significant funding."

"Yes, that was an unfortunate surprise." Oakes shook his head. "We need better information about what they're doing."

"Bribe someone?"

"No, no. If we approached someone and they reported it, it could endanger our funding."

Oakes hadn't offered him a seat, but Stenhouse sat down across from him, anyway. "I have an idea. There is someone I would trust not to betray us who might get inside."

51

THERE WAS A RECEPTIONIST at the entrance to Energy Unlimited's office building now, and she announced a visitor for Camila. "There's someone here to see you. Professor McGavock."

Camila muttered a mild expletive before answering. "All right. Send him in."

She blanked her screen, rose, and opened the office door. McGavock came down the hallway, and she greeted him coolly. "Good morning, Professor. What can I do for you?" She motioned him into a chair.

The teacher sat down and leaned over the desk. "I'm still very interested in your work. I know you have your own company now and are no longer connected to the school, but I would still like to help."

It wasn't easy for her to smile, but she managed it. "I'm not sure how you could, Professor. You probably know Professor Curtis is working with us because we needed an experienced mechanical engineer."

"Somebody told me that your original interest was in using Stenhouse technology to build spaceships. Is that true?"

Camila nodded slowly. "Yes, I hope to pursue that someday. Generating power seemed a better idea in the short term."

"I suppose I can't help much with energy production, but you might want advice on astrophysics when you turn to your ultimate goal. Have you really put it on the shelf, or do you give it some thought when you can?"

"I think about it." She almost said that her involvement in energy issues had become less now that the major problems were in engineering, not physics.

"Perhaps it would be productive if we talked about that occasionally."

The offer was tempting. *Would it hurt just to talk to him about space travel occasionally?* She could do that without telling him anything about EU's plans.

"Thank you, Professor. Sure, we could do that."

Two weeks later, Camila had the final paperwork for the patent application finished. Professor McGavock had called twice, but she put him off, saying that she was too busy. With others working on a design for a new prototype that would use vaporized fuel, she had more free time to think about spaceships. Investigating the physics of using a Stenhouse Field to generate power had given way to the engineering problems of applying the physics.

Applications for space travel were still only speculation, requiring more research. That was her area. When McGavock called a third time, she agreed to meet him in his office. Bringing Ben along seemed unnecessary.

Once they were settled in, McGavock opened the conversation. "If a generator expelled the positive ions and neutrons from a Stenhouse generator, the emissions would move very quickly, perhaps even near light speed. Using them to provide propulsion would give a specific impulse much higher than conventional rocket engines and more thrust than current ion engines."

Camila nodded. "Which means a ship could also reach much higher speeds. Not faster than light, of course, but still valuable for solar system exploration."

"So you're already thinking along those lines. What about down sides?"

"There will be plenty of issues to be addressed, but I think most of them will be engineering problems. One thing bothers me."

"What's that?"

"We call space a vacuum, but it really isn't. At the speeds that might be attainable, even the low particle density of space would be a problem."

"Yes, I suppose it would be." He paused and rubbed his chin. "If I knew a little more about your work, I might be able to help you with that."

That didn't take long. Camila smiled. "You already know the physics behind what we're doing. I'm afraid the engineering details have gotten beyond my ability to understand."

Professor McGavock shook his head. "I'm sure that's not true. You were one of my most promising students."

Where's a shovel when you need it? Camila held her smile, and they talked for another half hour before Camila claimed work to be done and left.

P ROFESSOR McGAVOCK SIGHED AND opened the requested meeting on his workstation. The screen blinked, and he was staring at the frowning faces of General Power Vice President Oakes and Lead Engineer Stenhouse. Their last conversation hadn't gone well, and McGavock didn't expect this one to be any better.

"What progress have you made?" Oakes asked. The executive was already frowning. He was not a patient man.

"The Lopez woman told me they are ready for another prototype test but was evasive about the changes they have made," McGavock said. "She mentioned spending much of her time on the patent application they have pending. She is implying that she doesn't know much about what they are doing, but I doubt that is true. I don't think she trusts me."

"We've developed a recovery design that doesn't violate their patent to deal with the reaction yields," Stenhouse said. "But we need more information. Are they still working on a solid fuel design?"

McGavock shrugged, but then hoped that the motion hadn't been noticed. "As far as I know. I believe they are working on increasing the energy output, but I can't be sure."

"You have told us they meet on Fridays to talk about their progress," Oakes said. "You should attend those meetings."

McGavock nodded. "I'll try."

52

CAMILA HAD PUT PROFESSOR McGavock off for two weeks, but after she finished her work on the Recovery Stage patent application, she had more time to think about future projects. Since that could involve space applications, the teacher's input could be valuable, and she relented. McGavock visited her at Energy Unlimited on a Wednesday in August.

Camila had been thinking about the problem of erosion on a spaceship exploiting Stenhouse technology. The astrophysicist could offer insight.

"It would be a problem for operation between solar system objects," McGavock told her. "Less of a problem for interstellar travel, of course, since you would typically leave the elliptic and encounter a more rarified environment."

"What about the particle density in the Oort cloud?" she asked.

"Well, yes. A ship going to another star would still have to deal with that. But it probably wouldn't be a problem. Larger objects are billions of miles apart." He rubbed his chin. "You would want better data on dust particles, though. They would be the real danger."

"Isn't the New Horizons probe going out there?"

McGavock smiled. "Yes, but it won't reach the Oort Cloud for hundreds of years. It doesn't have the instruments to study particle density at its current distance."

They talked for another twenty minutes before the professor changed the subject. "How is your company doing? Are you making progress on using the Stenhouse Field?"

Camila had wondered when they would get around to that. "I think so." She frowned. "I haven't been involved very much. The engineers have taken over. Professor Curtis was right when he said this was really more engineering than physics."

"Speaking of Curtis, Dean is still working with you, I understand."

"Yes. Technically, he's a consultant, but with his expertise, he's largely taken over."

"But you still go to those Friday meetings I've heard about."

"Sure, but a lot of it is over my head."

"I could probably help you more if I could sit in on one of those meetings occasionally. Are you having one this week?"

Camila hesitated. "I don't think the others would be all right with that. And we don't talk about my space travel ideas at those meetings."

"Still, unless you're talking secret design work, it shouldn't be a problem. And I might be able to help."

This is getting annoying. Camila smiled. "Let me talk to the others. You're right. The Friday meetings are high-level discussions of what we're doing, not deep dark secrets."

P ROFESSOR McGAVOCK WAS SURPRISED when Camila called him to say they would meet at the Energy Unlimited office building at six o'clock on Friday afternoon. He had a four o'clock class, so he had to hurry, but still got from Troy to Schenectady with time to spare.

Camila met him in the lobby and took him back to the conference room. Ryan Terry, Martina Patel, Gabrielle Pinto, and Ben Pinto were already there. Their greeting was friendly enough, but unenthusiastic.

"OK, let's get going," Ryan said after they were all seated. "I've got a date after this. Martina, how's the new design coming? And keep it high-level."

Martina nodded and glanced at her reader screen. "The new reaction chamber design looks good. According to the latest test results, we've solved the cooling problem. We're still limiting the lead core thickness to a tenth of an inch until we get the field control accurate enough to prevent a runaway reaction."

Ryan held up his hand. "Keep it high level, Martina. Gabby, what about control?"

Gabby shrugged. "You know what the problem is. The field configuration has to be adjusted almost instantly in response to the changes as the fuel is consumed. My team is getting there, but we're going to need more tests to nail down the parameters, and that means more quantum computer time."

Ryan frowned. "When?"

"We need another two weeks at least, assuming we can get the computer time we need to calibrate the calculations."

"General Power could already be working on this." Ryan's voice was harder. "We need to do better if we're going to have a product before they do."

Gabby glared at Ryan. "We're doing our best. We wouldn't be able to do it at all without my software."

She looked angry, and McGavock watched as her husband put a hand on her arm to calm her. The others looked at Ryan, who looked in need of calming himself.

"Fine," Ryan said. "We've talked about this before. Just get it done." He turned toward Camila.

"What about the patent application for the new design?"

"Coming along. I need to add the results from the latest test, but I should have the initial application ready within the week. Professor Curtis thinks the core design should get us another patent eventually."

McGavock saw Ryan's lips turn down slightly and thought he was suppressing another frown. *Because she mentioned a core again?* General Power was still working on using krypton as fuel, but it sounded as if EU was working with lead. Martina had mentioned the thickness of a core. Around a Stenhouse Field that controlled the reaction by constantly adjusting for the fuel left?

When the meeting broke up half an hour later, Ryan escorted McGavock out of the building. A taxi arrived less than a minute later to take him home, and as it sped away, he was already taking out his phone to call Zachary Oakes. He finally had the information he needed to get him in the good graces of General Power.

53

R YAN STROLLED BACK TO the conference room where the team members not meeting McGavock had joined the others.

"Do you think he believed us?" Camila asked.

Ryan shrugged and sat down. "I think so. You considered the fuel core idea for a while, so it should sound plausible to our friend, the professor. I don't know if Stenhouse will believe it, but they'll have to consider the possibility that we're doing something like that."

"General Power will have engineers on their team who might question it," Ben said.

"Maybe. I suspect, though, that they will underestimate us. At the least, they will think we're running with that and laugh at our foolishness." Ryan grinned. "If we're lucky, they'll decide they can do it faster than we can."

"It would be ironic if they were successful at it," Jürgen said.

"I think that's unlikely," Dean Curtis answered. "Not within any reasonable time, anyway. The control problem is too intractable." He turned toward Gabby and smiled. "Even with the algorithm you developed."

"We have the advantage there," Gabby said. "The algorithm and sample code are available in my thesis, but turning it into operational code is something else." She grinned and opened her hands on the table. "And they don't have my brother's program and the world's leading expert in integrating my algorithm."

Ryan smiled as well. "OK, so much for Professor McGavock. The pizza will be here shortly, so let's talk about what we're really doing before we stuff ourselves. First, I have news. Basu is sending their two board representatives next week to look us over: Amrita Savant and Sahil Nanda."

"Her again?" Jürgen shook his head. "Stick close by me, Cam."

"So, what are we going to tell them?" Ryan asked.

"That could be delicate," Dean answered. "They're owners and are entitled to know how we're doing, but we don't want to reveal too much."

"I know Amrita could be a little, ah, unethical in pursuing an objective," Ryan said. "But you don't think we can trust them?"

Dean shrugged. "I think if we're successful, the government will be concerned about leaking our technology to other countries. It will be better for us if we can show we haven't done that."

Ryan nodded. "Maybe you're right. Okay, so let's talk about where we are and then we can decide how much to tell Basu."

Dean entered a command into his reader and glanced at the screen. "The test of the first stage went well. The lead fuel vaporized and fed smoothly into the reaction chamber. We're still using the original chamber design, but I think we can improve it.

"We're using the electrical grid to initiate the Stenhouse Field and power the first stage right now. Ultimately, we will need an external source for the field, but if we use the idea of injecting a bit of krypton into the field, we can feed that energy back to the first stage to start vaporizing the fuel. After that, it would be self-sustaining."

"How much of that is patentable?" Martina asked.

Ryan frowned. Martina was an Indian citizen, in the United States with India's support. *Is anything they were doing really a secret from Basu?* He nodded to Dean.

Dean faced Martina. "I think our design for the first stage will be. Because of the materials we'll have to use, the software, and the nozzle complexity, it should be unique enough to be protected."

Martina looked at Ryan as if she were reading his mind. "I'm not passing on any information, but it will be public once the patent is issued."

Ryan shook his head. "That's true of the pending patent for the recovery stage, but our patent attorney has suggested that we apply to the Commissioner of Patents to have additional patents declared secret."

Martina sighed. "Maybe I should leave. While I'm working at the machine shop, I see little of the design, anyway."

"We won't be discussing any sensitive details," Camila protested. She looked at Dean. "Will we?"

"I had only planned to tell everyone our status." Dean smiled. "If I went into the engineering details, I think some of you would get lost pretty quickly."

"Then we don't have a problem," Ryan said. "We can give Basu the same high level status information that Dean is giving us now. They don't need to see our designs. At least not if we can make the patents secret."

T HE BOARD MEETING ON Wednesday of the following week was a rollercoaster. It began on a friendly enough note when Amrita and Sahil arrived. Only Ryan, Camila, and Ben met with them. The other two board members, Gabby and Jürgen, opted to work at V-World.

"We'll all get together for dinner tonight," Ryan promised. "It will be quite a crowd, though. We're growing."

Over the next half hour, Ben presented the status of their research and summarized the data obtained from the prototype. Finally, he briefed them on the design proposed for the next prototype.

"All the components I described will be included," Ben said in his conclusion. "If we are successful there, we plan to begin design on a small Stenhouse reactor that we can sell."

"When can we expect such a product to be released?" Sahil asked.

"It's difficult to make an estimate before we have a design, so you shouldn't take anything I say too seriously," Ben answered. "We think we can manufacture the first models using our existing facilities in about two years. Large-scale production will require a new, larger factory."

"We have begun negotiations for a tract of land not far from here," Ryan added. "Using your investment, we expect to break ground in the spring."

Sahil's eyebrows went up. "Not until then?"

"It's possible some work might be done if negotiations are completed soon," Ryan said. "But winters here make construction difficult, and we plan to use the time to better understand what we will need. Our design progress may affect some aspects of a new facility."

Sahil nodded. "We look forward to seeing your designs."

Ryan hesitated and cleared his throat. "Well, I'm not sure we'll be able to show them. Not all of them, anyway."

Sahil frowned. "Explain, please. I represent an owner of this company."

And, for a while, the meeting was much less friendly. Ryan tried to explain about the probability that the United States government would have security concerns which Energy Unlimited planned to assuage by applying for secret patents for future designs. It was unlikely that the government would want those designs available outside the United States.

Sahil was not happy about that despite their earlier agreements about access to the technology. He protested vehemently. Amrita was quiet, although Ryan could see she was disturbed as well. She had introduced Sahil as her superior, so her silence was not surprising.

Eventually, Sahil seemed to accept the situation, and the meeting turned to a less acrimonious discussion of plans for the availability of products and the distribution of profits. By late afternoon, they were more than ready for a relaxing dinner.

T HE DINNER HAD BEEN pleasant, but a feeling of unease tempered Martina's enjoyment. She couldn't forget that she was still an Indian citizen, dependent to some extent on her government to remain in the United States working for Energy Unlimited. She had planned to share a ride back to Troy with Rafael Rodrigues, their materials engineer, but as she walked to the front door, Amrita beckoned to her.

That didn't help her discomfort, but she turned to Rafael. "I think I'm being summoned. Maybe you should go back without me."

"I'll wait if you want me to."

Martina smiled. "Thanks, but don't wait too long. If I'm not back in a couple of minutes, I will probably be a while."

Rafael nodded, and she walked toward Amrita. Amrita moved into a corner where, presumably, they would be less noticeable. Martina looked back toward Rafael, but Ben and Gabby had joined him and had his attention.

"I've called for a taxi," Amrita said when Martina was close. She turned so that her face was hidden from the group around Rafael. "I'll bring you back to your apartment and we can talk in private on the way."

Martina nodded but didn't speak. She could guess what Amrita wanted to talk about. Her earlier discomfort was growing into almost painful stress, but she followed Amrita outside where a taxi waited.

"I just wanted to tell you that your country is behind you," Amrita said when the taxi moved away from the restaurant. "I know you also will serve your country."

Martina stared at Amrita but still didn't speak.

Amrita smiled. "You're shocked. You think I want you to spy on your friends." She put a hand on Martina's shoulder. "Surely you know Basu Power is entitled to know everything about what Energy Unlimited is doing. We're part owners, not some outside interest."

I have to say something. But it all came out in a rush of words. "I don't really know anything. I work at the machine shop, and I see the manufactured parts, but I have no access to control circuits or software. That's where the important design details are."

Amrita looked at Martina and frowned. "I'm sure you can help us if you try. Just remember that we're entitled to everything as owners."

That isn't what Ryan said. Ryan's lawyers had drawn up the agreement between Basu Power and Energy Unlimited, but Martina didn't know what it said. She knew nothing about the legal issues, either.

Martina felt loyalty to India. Her family still lived there, and she owed her country for her education and almost everything else. But she also felt loyalty to the friends she had made and especially to Mary Welch, the lonely woman struggling to survive after losing her husband.

When the taxi arrived at her apartment, she gave Amrita a vague commitment to see what she could do and hurried inside. She had to think.

54

Energy Unlimited was busy. Dean Curtis, once a consultant and now an official employee, hired four more engineers to help him design a small generator suitable for commercial use. Prototyping the major components followed. Ryan shepherded the purchase of land for a factory and contracted a construction company to do site preparation.

Marlee Dubois and Emma Martin, the electrical engineers EU had hired in January, were tasked with developing the control circuitry for the generator, and added two more engineers to their department. Gabby moved from V-World to EU to write the software, warning Ryan that she would need more people soon as well.

The AI company that purchased V-World approached Jürgen for information about Energy Unlimited. Their software required immense server farms and prodigious amounts of electricity, and Stenhouse generators were a promising alternative to existing sources.

Senator Leo Correia entered his home in a jubilant mood. That afternoon, the chairman of his party had told him his name was being mentioned as a candidate in the next presidential primary.

"People like you," the chairman had told him. "I think you're the man who can restore our party's position and preserve our nation." Correia looked forward to telling Priscilla.

But his wife was less enthusiastic than he expected. "This is because of your statements about homosexuals," she said. "It's nonsense, and you know it."

His cheerful mood became indignant annoyance instantly. "I know no such thing. I'm popular with most people, but gays aren't doing their part for the country."

"In that one respect," Priscilla answered. "They work as hard as anyone on everything else. And they used to adopt orphaned children before the government made it so difficult. How much money would that save local governments?"

"We shouldn't endanger our youth just to save a few dollars."

Priscilla glared at him and moved closer. "You'll support that nonsense, harass gays, and expose them to every bigoted creep in the country to get a few votes?"

"It's what I have to do to win elections." Correia shook his head and stalked away.

"Oh, it's not money, it's winning elections," Priscilla shouted after him. "That makes it all okay."

In his office, he asked his computer for the Senate News Report. The screen blinked twice, trying to interpret his disgruntled muttering, but then settled on the requested information.

There wasn't much that he didn't already know. An item on General Power's request for government support caught his eye briefly, but only because he remembered interceding on their behalf. *That must have been a year ago.* It wasn't something his party wanted to deal with, but the opposition was keeping the issue alive. He could make a statement criticizing their inability to get it done, but that would imply it should get done. If the leadership of his party disagreed, it could dampen enthusiasm for his candidacy.

He leaned back in his chair and stared glumly at the computer screen. His wife just didn't understand how his world worked. Still, he would have to find a way to placate her.

ARTHUR STENHOUSE KEPT HIS expression carefully neutral as he watched Zachary Oakes take his seat at the head of the conference table. Stenhouse didn't doubt that the man was knowledgeable about conventional power generation. He wouldn't have progressed as high as he had otherwise. But the advanced physics that Stenhouse had studied for decades was beyond a mere engineer.

Stenhouse addressed everyone, but kept his gaze on Oakes. "I believe we are well ahead of Energy Unlimited in developing the solid core Stenhouse generator."

"Has your professor told you that?" Oakes asked.

"He hasn't learned much lately. But, given the challenges we're facing, they can't be making much progress."

Oakes frowned. "I've been told that they've purchased land and are ready to build a manufacturing facility. Did you know that?"

"I can only surmise that the investment from India has made them overly optimistic. They have also been hiring more people, but not enough with the skills to surmount the control difficulties we're dealing with."

"How can you be sure of that?"

Stenhouse smiled. "By now, they probably have a crude prototype, but, as I'm sure Ian will tell you, initiating a low-power reaction is one thing. Controlling it over an extended period is more difficult."

Ian Winston, heading a team of ten engineers designing the generator controls, nodded. "We're finding it exceedingly difficult. Constantly adjusting the field to maintain energy output as the fuel configuration changes requires extremely fast circuits. That isn't easy for such a complex task."

Oakes stared at Stenhouse. "We're putting a lot of money into this. Can you get this done and have a product before Energy Unlimited?"

Stenhouse raised his head and looked back at Oakes confidently. "Of course. I discovered the Stenhouse Field. I have no doubt I can exploit it before those amateurs."

Oakes smiled. Stenhouse smiled back, sure he had convinced Oakes he could make their project a success. *I just have to make these engineers do their jobs.*

55

J ÜRGEN CAME TO THE EU office to take Camila out to lunch. They sat in the coffee shop a block from the EU office, talking while they nibbled on sandwiches.

"I've been spending most of my time going through the data from the prototype tests," Camila said. "Not very exciting, but I might find insights that will help us some day."

Jürgen grinned. "When you turn to designing spaceships?"

Camila smiled back. "Of course." She stopped and looked toward the door where Ben was entering and waved him over. Ben nodded and ordered a sandwich at the counter before joining them.

"Cam was telling me she doesn't have enough to do," Jürgen said.

Camila glared at him. "That's not what I said."

Ben's smile was half-hearted. "That's probably what you meant, though. I feel the same way."

"Why?" Jürgen asked. "According to my sister, things are busy. She doesn't have time to help me anymore."

Ben shrugged. "Oh, EU is very busy. For the engineers, anyway. Us physicists aren't needed much now."

Jürgen looked at Camila. "You do seem to have lost some of your enthusiasm. You're more tired now when you come home, too. I thought you were working too hard."

"More like boredom," Camila said. "I'm not sure what I'm doing there."

"You're not thinking about looking for a new job, are you?" Jürgen asked.

Camila shook her head. "No. I'm an owner, and they do need to consult with Ben or me occasionally."

"Same here," Ben said.

Jürgen considered that. "I have an idea. Cut back on your time at the office. Work half days, maybe."

Camila frowned. "How would that help? What would I do with the rest of the day?"

Jürgen grinned. "You still want to build better spaceships someday, right?"

"Sure, but we're not ready for that yet."

Ben's face lit up. "I think I know what you're leading up to. You think Camila should go back to school."

"Exactly." Jürgen leaned forward. "You don't even have to go for another degree. Take courses that might be useful when you do build spaceships. Astrophysics, engineering, maybe even more physics courses. Chemistry might be useful too."

Why haven't I thought about that? Camila felt more excited than she had in months. "You're right. You too, Ben?"

"I don't see why not. I miss the college sometimes, and it could help in the long run."

"Astrophysics." Camila frowned. "I took an undergraduate course, but advanced courses are available."

"You sound reluctant," Jürgen said.

"Professor McGavock teaches astrophysics courses."

"You could take the courses remotely, couldn't you?" Jürgen said.

"I could, but I miss the college sometimes too."

"Don't worry about astrophysics, Camila," Ben said. "I wouldn't mind learning more. And I can handle the professor."

"I thought you were more interested in quantum physics," Camila said.

"I was. Actually going into space has changed my perspective, though. I think I'm coming around to your point of view. Building spaceships sounds appealing."

Camila jerked her head up and down. "Then let's do it."

"Registration for the spring has ended, though," Ben said.

"Talk to Professor Moreno," Jürgen suggested. "She might be able to get you an exception."

P ROFESSOR MORENO WAS DELIGHTED to intercede for Camila and Ben and register both for courses in the spring. Camila enrolled in an aerospace course in propulsion and an advanced quantum gravity physics course. Ben had already taken the quantum gravity course and decided on a mechanical engineering course on cooling system design. He also signed up for the advanced astrophysics course.

CAMILA WAS WALKING TOWARD the classroom exit after her second session in the propulsion course when a student intercepted her. "Hi, I'm Julie. You're Camila Lopez, right?"

"Hi. That's me."

Julie was almost jumping up and down with excitement. "You're a legend among us undergraduates. I can't believe you're in this class with me."

Camila smiled. Julie was a few inches shorter and undeniably cute. Her flawless chestnut brown complexion almost glowed and her curly black hair perfectly framed her round face. Camila's realization that she was attracted to the girl was disturbing. *How did I not know before?*

But what's this about being a legend? "What do you mean?"

Julie seemed to get control of herself. "You started your own company less than a year after graduation. We know all about you."

Camila held up a hand. "I had a lot of help and some luck. And we don't know if we're going to be successful."

"I'm sure you are." Julie hesitated and her expression softened. "I would love to hear about it. Maybe I could buy you a coffee at the Student Union?"

Camila had planned to go back to her office, but she had no reason to rush. It was early for lunch, but coffee sounded good. "Why not? But I'm buying. I remember what undergraduate student finances were like."

They didn't talk much on the walk to the Student Union. Winters were usually milder than they had once been, but temperatures in upstate New York were still cold, and a stiff wind made it worse, deteriorating their voices even more than the masks they still wore.

The Student Union was a large, two-story building with offices and several facilities, including The Canteen, a small restaurant with a wide selection of coffees. It was a student favorite, especially in the morning, but they found a table and removed their gloves, jackets, and masks. After getting coffee, they sat down.

For the next ninety minutes, Julie peppered Camila with questions about her and Energy Unlimited. Camila noticed other students staring at them, women as well as men, but didn't think about it much. Julie was more interesting. But before parting, Camila agreed to meet her again after their next class together.

The following week, they met again at the Student Union, drawing stares again, and many were hostile. Other tables with two or more women and no men didn't seem to draw the same attention. Camila enjoyed their meetings, but wondered if they should find some place more private to talk.

This time, before Julie left for another class, she pulled Camila into a hug. It felt good, and when Julie pulled back, she smiled and touched Camila's cheek. "Next time."

Camila watched her stride away and tried to analyze her feelings. She had been immediately attracted to Julie, but had thought Julie's interest in her was because of her success with Energy Unlimited. Now she wondered if that was true.

AFTER GRADUATING WITH A degree in architecture, Greg Mora went to work for a construction firm in Latham, a city across the Hudson River from the college. He still lived in Troy, rooming with an architecture student who had interned with the firm over the summer. Bob Gamble was two years younger than Greg, but they had quickly become friends.

"I saw your old girlfriend today," Gamble told him when he got back to their apartment.

Greg had told him about Camila and showed him pictures of her. In his stories, he was the one who had ended the relationship because Camila had been, in his words, "a cold bitch." Gamble's comment grabbed his attention, but he tried to appear indifferent.

"Camila? Is she still around?"

"I saw her with another girl in the Student Union. I think she's still taking classes."

"She started her own company. Has that failed already?"

Gamble shrugged. "I don't know. But she was with Julieta Williams. I thought you would find that interesting."

Greg waved a hand. "Why would I care?"

"Williams is active with a gay group on campus. They seemed pretty friendly."

"That would explain why she was so cold with me. I suspected she was a lesbian."

"The group is meeting tomorrow night. Maybe we should check it out."

Greg hesitated. "It will be cold."

Gamble grinned. "Maybe Camila will be there."

Greg returned the grin. "All right. I'm in."

56

Camila hadn't mentioned Julie to Jürgen after the first meeting. When she got home after her second coffee date and he asked how her day had been, it all came out in a rush.

"I'm glad you found a new friend." But Jürgen didn't look happy. He confirmed that when he added, "Just be careful."

Camila frowned. "Some people looked like they disapproved. Is it that bad? Two women can't even have coffee together?"

"The reaction is unusual. Just be careful."

She saw Julie again two days later after another propulsion class, but Energy Unlimited was having a status meeting that afternoon and Camila needed to attend. She talked to Julie briefly and made tentative plans to get together on the weekend. Then she hurried back to Schenectady.

The Energy Unlimited conference was encouraging. Dean reported substantial progress on a design for a small Stenhouse reactor that could become their first product. He expected its energy output would be sufficient to run a moderate-sized factory, and Martina suggested it could also find a place on large ships or locomotives. It would need heavy radiation shielding, making it expensive for smaller purposes.

The usual Friday pizza fest the next day was more raucous than normal. Beer was consumed in greater quantity than usual, and Camila and Jürgen were both feeling it on the short ride home. Both were a little unsteady as they navigated the sidewalk and fumbled their way in.

"I'll see you in the morning," Camila said, moving toward her bedroom.

Jürgen was looking at a blinking light on the television. "I'm just going to check this notification."

The television blared into life. "Anti-gay demonstration turns violent. Do you want more details?"

Camila turned as Jürgen swore and answered affirmatively.

"A demonstration outside the meeting place of the college homosexual group Pride Alliance turned violent tonight as attendees left the meeting. It is uncertain at this time what triggered the incident, but three people were injured, one seriously."

Jürgen glared at the screen, showing a small crowd and two police cars. "Uncertain? Like hell."

"Hospitalized was Julieta Williams," the television said. It went on to name two other people released with minor injuries.

Camila moved closer to the television and stared at the display showing Julieta Williams. "That's Julie! She's in the hospital?"

Jürgen nodded. "They're not saying how badly she was hurt."

"I should go." But she moved unsteadily to a couch and collapsed onto it.

The report continued. "Police questioned two men about the incident but did not make an arrest. Their names have not been released."

Camila looked at the screen again where the two men questioned were shown talking to a police officer. She recognized one of them. *Greg Mora!* She hadn't seen him since the night Dani tripped him, and assumed he moved on after graduating. *Was he one of the attackers?*

"I DON'T THINK THAT'S a good idea." Jürgen wasn't sure he was thinking clearly about everything, but he was sure about that. He understood the hostility in the coffee shop Camila had reported. If her friend was open enough about her homosexuality to attend meetings of a campus gay group, bigots would assume a woman socializing with her was also gay.

Camila sat on the edge of the couch, still staring at the television. "What? Why? Because I'm a little drunk?"

Jürgen shook his head. "Because you might be identified, and people will know you're gay too. And, since you're living with me, they'll wonder about me as well." Camila turned toward him, and he could see fear in her eyes.

"Oh, god." She fell back against the couch and covered her face with her hands. "What can I do?"

"Maybe she's not hurt too badly. We can find out in the morning."

"The report said she was serious."

"We'll find out in the morning."

S HE WAS TIRED AND still feeling the effects of the alcohol, but Camila couldn't sleep. Knowing that Julie was hurt but not knowing how bad was stressful enough. But it didn't end there. Julie hadn't told Camila she was gay. They didn't know each other well enough for that, and Julie was probably uncertain about Camila's orientation. Still, others knew about Julie and would draw their own conclusions.

Jürgen had warned her and even told her about his former lover. Camila hadn't listened, thinking that was an isolated instance, but Jürgen was right. There were people who would hate her just because of what she was, dangerous people.

If someone recognized her, not that difficult, they could come after her too, and that would lead them to Jürgen. Their bigotry might even extend to others they worked with, regardless of whether they too were gay.

Jürgen hadn't wanted her to go to the hospital right away, implying it was too soon. But in the morning, he would advise her not to see Julie at all. Not at the hospital or anywhere else where someone might see them.

She told herself she hardly knew Julie. Maybe a few coffee dates wouldn't have become any more than that. That wasn't the point, though. How could she ever have any relationship beyond friendship? And was even that safe?

The same thoughts revolved around her brain in a confused loop. Tension knotted her stomach, demanding release, and she allowed a few tears to trickle across her cheeks. Like a cracked dam, she burst then, sobbing into her pillow and hoping Jürgen wouldn't hear.

57

W HEN CAMILA CAME INTO the kitchen the next morning, Jürgen was already there, staring at a cup of coffee. "Your friend has a concussion and a couple of broken ribs, but she's not critical," he told her.

Camila nodded, but couldn't find anything to say. She poured a cup of coffee and sat down at the table. After a minute of silence, she finally found a word. "Sorry."

Jürgen glanced at her and then turned his gaze back to his coffee cup. "For what?"

"For putting you in danger. For putting all of us in danger."

"It's not you." His voice was a growl. "It's all those assholes." He got up and stalked over to the glass door leading to his backyard patio. He stared out, but if he was looking for someone, there was no one there. "Damn it, why can't they just leave us alone?"

"I can't see her again, can I?"

Jürgen turned, his face a fierce mask. Camila hadn't seen him that angry before, but she wasn't afraid. Not of him.

His face softened. "Not in public. I don't think you should."

She had taken her phone with her, and it announced a call. Camila didn't recognize the caller, but spam calls had become extremely rare over the last couple of years. She answered.

"Good morning. This is Katherine Lovejoy from WTPR Media. Is this Camila Lopez?"

Jürgen could hear the conversation, and his face darkened again. Camila took a deep, shaky breath before answering. "It is."

"Are you aware of the incident at the college group meeting last night? And that a girl was injured?"

"Yes, I saw it on the news."

"You weren't there?"

"No, I was with work friends until fairly late."

"The person injured was Julieta Williams. Did you know her?"

Jürgen was shaking his head, but she couldn't deny it. If this reporter already had her name, people who saw them together had identified her. "Slightly. We're taking a class together."

"That's all?"

"We had coffee together after class a couple of times."

"Nothing else?"

"What else would there be? We only met last week."

"Do you know Miss Williams is a homosexual?"

Camila wanted to shout at the woman, but she kept control. "I gathered that when I heard where she was injured."

"So you didn't know before that?"

"We only had coffee together. The subject didn't come up."

So far, she had only told the truth, but the reporter had one more question.

"Are you a homosexual?"

Camila didn't think her hesitation would be noticeable. "No."

J ULIE DIDN'T COME TO classes for the next two weeks, and Camila couldn't get any information from the hospital about her condition. Jürgen had shown her a news report two days after the attack saying she was recovering, but after that, there was nothing.

Would she even come to class again? The experience must have been traumatic, and she might be concerned about how people would treat her upon her return. Camila was sensitive to other people when she was on campus, but realized after the first week that she wasn't getting any extra attention. Would it be the same for Julie?

Camila didn't know what she would do if Julie returned. Should she pretend she didn't know Julie, as Jürgen suggested? That was the prudent choice, but was it the right one? Then, eighteen days after Julie's injury, Camila walked into the classroom and saw Julie sitting in a back row, looking like she was trying not to be noticed.

She had no outward signs of her injuries, but the joy that had enhanced her attractiveness was gone, replaced by a sad wariness. She glanced toward Camila from across the room but didn't smile. Camila nodded before she took her usual seat. She wasn't sure Julie noticed.

After class, she joined the other students drifting out. Julie still sat in the corner, looking down at her lap, and Camila hesitated before letting the stream of people carry her into the hall. Outside the building, she walked away, pulling her coat tight against the February cold and a biting wind.

J ÜRGEN HAD STARTED A batch of chili in his slow cooker that morning, and the hot, spicy bowls he served for dinner were perfect for the freezing weather. Camila ate slowly, eyes staring at the food but probably not seeing it. He doubted she tasted it, either. That morning, she had seemed to be regaining her usual cheerful mood, but the day had set her back.

He debated asking her about it and dithered until they were clearing the dishes. "How was your day?"

Not even a hint of a smile. "It was okay. The usual."

"Really? That's not what your face is saying."

She turned to look at him with watery eyes. "Julie came back today."

"That's good, isn't it? She's recovered."

Camila looked at the floor. "She sat in the back of the room, looking miserable. I didn't talk to her. I wanted to, but I was too scared."

What should I do? He didn't know what to say and didn't think hugging her would be the right thing to do. He compromised and put a hand on her shoulder. Camila looked up at him again, trying to smile. She didn't succeed, but only moaned and left the room. He heard her bedroom door close and stood in the kitchen for a while. *How can I help her?*

C AMILA ENTERED THE CLASSROOM for the last session of her propulsion course. It had been four months since she met Julie and the attack ended the budding relationship. Julie still attended the class and still sat in the back of the classroom.

Inevitably, there had been times when their eyes met, or they were in proximity. They nodded to each other but didn't speak. Camila knew she should say something, but she didn't have the words. *Or the courage.*

As she walked toward her usual seat, she looked back toward the entrance as another student entered, but it was only to steal a glance at Julie. She was looking down at the table, but must have sensed Camila's gaze. She looked up and smiled, but it was a sad smile. Camila returned the smile but doubted it was any happier.

There was a folded piece of paper with her name on it on the table at her seat. She opened it, but there were only two words and no signature. "I understand."

58

W HEN CAMILA WALKED INTO the conference room for the usual Friday night pizza fest, almost everyone else was already there, and they were all in an exuberant mood. It was the middle of July, and a new prototype had been successfully tested earlier. Most of the team was sitting at the table, but Dean and Mary stood next to the network screen where the prototype design was displayed. As Camila took a seat, Dean was pointing something out to Mary on the screen.

"We've analyzed the test results," Ben said. "Performance is everything we hoped for."

"Maybe even better than we hoped for," Dean said. "We'll need to add shielding before we can test it at higher fuel rates, but it already turns a tiny pellet of lead into enough energy to power this building and the rest of the block for a month."

"They expect me to manufacture that shielding ASAP," Mary Welch said. "Like I don't already have more to do than I can manage."

Dean put his arm around Mary. "We know we can count on you."

"Once our manufacturing plant is finished, you can get back to your paying customers," Ryan said.

Mary scowled at Dean, but she didn't push him away. "When is that?"

Dani was managing the construction for Energy Unlimited. "The contractors say September. We'll be able to work on the interior over the winter and be done by June, but Dean plans to use our first working generator to power the facility. If that's not ready, it could cause a delay."

"Given the results today, and," and Dean paused and looked down at Mary, "continued cooperation from our favorite machinist, we should have a first unit by March."

"Sahil has already told me he wants an early unit for Basu," Ryan said. "For demonstrations, I assume. He has also asked about the possibilities of a full-blown generating facility."

Dean nodded. "That will take a little longer. Something on that scale will require more than just exploiting the technology."

"He's talked to me also," Martina said. "I told him it would probably be years before we could manage a project on the size he wants."

Ryan raised the glass of beer he had been drinking. "But we'll do it."

B EN APPROACHED CAMILA AND Jürgen as they prepared to head home. "Can I talk to you, Cam." He motioned them back to the table, and they sat down. Camila looked at him expectantly.

"If we're going to take more classes in the fall, we need to register," he said. "Do you know what you're going to take?"

"Not yet. I was going to see what's offered this weekend."

"The school has added a graduate course on the Stenhouse Field. I'm going to sign up for it."

"At this point, I would think you could teach it," Camila said.

"Maybe. But Professor McGavock will be the instructor. I think it could be useful to hear what he has to say."

"Insight into what General Power is doing?" Jürgen said.

"Sure. But I might learn something about the physics involved, too."

"I don't think I want to deal with him," Camila said. "It's not really his field, anyway."

Ben nodded. "I'll have access to the class videos, and I can play them for you if he says anything interesting."

Camila seemed to think about that, and her next statement confirmed it. "That would work. I'm not going to go to the campus for the next semester, anyway. I waste too much time traveling between Troy and Schenectady."

Ben nodded as if he accepted her decision and the reason for it, but she had put the travel time to productive use. Although Camila didn't talk about Julie, he was sure what had happened to her friend motivated Camila's decision.

B EN HAD HIS OWN reservations about meeting his old mentor again, but Professor McGavock noticed him enter the classroom and greeted him warmly. Ben returned the greeting and took a seat.

At the hour, McGavock welcomed the students and introduced himself. Then he glanced at the notes on his podium and continued.

"Twelve years ago, Arthur Stenhouse made an astonishing discovery that we are only now beginning to understand. The Stenhouse Field was unsuspected until his work then, but it presents us with unparalleled opportunity." He paused and moved a few feet from the podium.

"For astrophysicists like me, it is a clue that may explain the mystery of dark matter. It may also help scientists develop the ultimate prize, unification of general relativity and quantum mechanics. That was enough to fuel my interest in the field and push for this course."

He was getting warmed up, and paced across the front of the room. "All that is wonderful for students of physics, but doesn't mean much to the common man on the street. However, Doctor Stenhouse is continuing his work with General Power, work that I have consulted on. Their goal is to exploit the Stenhouse Field to provide essentially unlimited energy by fissioning non-radioactive materials. His project has already been successful using the noble gas krypton as a fuel."

After Camila did. Ben wanted to interrupt and point that out, but he held back.

"This will be tremendously important to that man on the street," McGavock continued. "We are seeing extremely encouraging efficiency from prototypes, suggesting lower costs than current generating methods. As an example, converting a small amount of lead into iron could power a small city."

We? How much help is he really giving Stenhouse? Ben shook his head. *Was the course going to be one long advertisement for General Power?*

McGavock returned to the podium and glanced at his notes again. "But all that is just to tell you how important this topic will be in the decades to come. The goal of this class is to teach you all the basics of the Stenhouse Field and how it fits into physics as a whole."

Mostly, he went on to do just that. He did insert an occasional reference to General Power, but that usually fit with the material. Ben found it interesting that he had only mentioned the original work with krypton and not the current—troubled?—work with solid fuel.

McGavock pointedly did not mention Energy Unlimited or Camila's thesis work, the basis for everything that General Power and Energy Unlimited were doing. Even with Ben in the class, he didn't point out that Ben's research had confirmed Stenhouse's claims. But Ben resisted bringing that up, concerned that he might reveal what they were doing, as opposed to what General Power was supposed to believe they were doing. If he divulged nothing directly, there still might be questions, and even no answer could reveal something.

At the end of the class, McGavock looked at him with eyebrows uplifted in question. Ben realized the teacher had expected him to interrupt. *Dodged that trap.* He returned the look with a grin before leaving the classroom.

59

IT WAS THE MIDDLE of March, and spring was over, judging by the temperature, if not the calendar. The first generator was complete, and Ryan suggested the usual Friday night meeting should be special.

The generator was moved from the assembly room where the prototypes had been built to what would be the power room in the factory still under construction. Ryan rented tables and chairs for an adjacent room and arranged for a catered dinner. Sahil and Amrita came from India, adding to the twenty people now employed by Energy Unlimited.

Ryan stood near the warming table where the food was ready to be served. He glanced at the table and then back to the room and the attendees waiting expectantly. "We could eat now. But the food will stay warm for a while. Perhaps you would all prefer to see Energy Unlimited's first product."

The enthusiastic response was everything he expected. Only Dean Curtis and his team had seen the completed generator, although most had viewed parts or design documents. Ryan grinned and shrugged. "Okay, if that's what you want to do. Dean, why don't you lead the tour?"

Dean rose and stepped toward the door. "It would be my pleasure, Ryan. Everyone, please follow me."

He led them to a nearby room that was too small for the number of people that crowded in, especially since they tried not to get too close to the massive black cylinder in the center of the room. The thick cooling fins projecting from the main body made that tricky.

"The generator isn't operating yet." Dean had to shout to be heard over the exchange of comments as everyone talked at once. "Even if it were, it would be perfectly safe, although the room might get uncomfortably warm."

He patted a fin and ran his eyes up and down the generator's ten foot height. "The outer shell is four inches of lead shielding. A fair amount of the generated energy is released as radiation. A high-speed stream of positive ions—protons, iron, and nickel chiefly—creates a magnetic field that generates electricity. Water flows through the casing, cools the generator, and circulates through these large fins before being sent back into the main casing, but the flow could be harnessed to drive a turbine."

"The fuel is lead, right?" Jürgen said. "How much?"

"Technically, it can run on anything with a molecular weight higher than iron," Dean answered. "But we plan to use lead, and estimate that we'll need about a pound per hour to run the factory when it's fully operational."

Dean moved to the other side and opened a heavy door atop a section extruding from the cylinder. "We load lead pellets in here. A pre-stage liquifies the lead and feeds it into the first stage of the generator. It vaporizes the fuel and injects it into the reaction chamber."

Ryan scratched his head and wondered if he was about to ask a dumb question. "Which came first? The chicken or the egg?"

Dean grinned. "When the generator is first started, it has to establish the Stenhouse Field from an outside source. This unit uses the normal electrical grid, but a battery is a possibility too. Once the field is on, a small amount of krypton is injected to start the reaction. That releases enough energy to begin a self-sustaining process."

"You're using this unit to run your factory," Sahil said. "But the factory can't manufacture the units in quantity until it is completed. How soon before more units are available?"

There, at least, Ryan was on firmer ground. "As you point out, the factory isn't ready yet, so additional units will have to be manufactured by hand with parts from Mary's operation. I know that you have requested the next two units for Basu Power, and the first should be ready to ship in about two weeks. Another two weeks for the second."

"We'll have the specifications ready to give you before you leave," Dean added. "Power output, connection requirements, and so on."

Aɴᴏᴛʜᴇʀ Mᴏɴᴅᴀʏ ᴍᴏʀɴɪɴɢ. Aʀᴛʜᴜʀ Stenhouse vaguely remembered a time when he looked forward to the first workday of the week. He hadn't felt that way in a while.

Two years had gone by since he promised Zachary Oakes that they could beat the Energy Unlimited upstarts to develop a practical Stenhouse generator. They designed a way to manage the output from the reaction that didn't violate EU's patent. That hadn't been a problem, only a delay. He had overseen the building of a prototype using their version and a reactor using a lead core fuel, the same principle as McGavock had said EU was using. Getting beyond the prototype still presented intractable problems in controlling the reaction. Secondary, but still important, was the issue of refueling a design with a solid fuel core built into the reaction chamber.

Several of the project engineers had suggested alternatives, including the krypton fuel design first used by Camila Lopez for her thesis. He had opposed throwing away the work they had done, and Oakes had agreed. At least Energy Unlimited hadn't succeeded yet.

Two weeks before, there had been inquiries from above Oakes when Energy Unlimited filed another patent application. That had caused some consternation, especially since the Commission of Patents had allowed the patent details to be kept secret. The application was for a gas injector device, but didn't contain more details. Gas injectors were ancient technology, so it was a mystery how the application could show uniqueness, but it suggested the competition was going back to a krypton-injection design.

The blowback from that was finally fading and Oakes had stopped visiting him daily to harangue him about their progress. It had been difficult to convince Oakes it wasn't Stenhouse's fault if the engineers on his team weren't up to the task.

He couldn't suppress a groan when he saw Oakes waiting for him as he approached his office. The expression on the executive's face left no hope that he came with good news.

"Have you seen the news?" Oakes asked with an agitated wave of his hands.

"Just the world headlines." Stenhouse realized too late that Oakes probably meant technology updates, and Stenhouse rarely checked those until he was in his office, and seldom right away.

He reached the door to his office, and Oakes opened it and pushed him inside, slamming the door closed behind them. "Energy Unlimited is in the news. I'm surprised the world news didn't cover the story."

Suddenly, his knees wouldn't hold him, and Stenhouse dropped into his desk chair. Only one thing would get the attention Oakes was talking about, but he asked anyway. "What about Energy Unlimited?"

"According to their release this morning, they have a working generator powerful enough to supply the factory they're building in New York. They're claiming a few pounds of lead will be enough fuel for a day of heavy manufacturing."

Stenhouse shook his head. "They've solved the control problem, but how are they going to refuel it? How often will they have to replace the core?"

That question arose again at a meeting an hour later, attended by the CEO and one of the senior project engineers. Neither looked happy.

"They're not replacing any core," Sam Householder, the engineer, said. "They never planned to use that design."

"What do you know about that?" Paulina Alvarez, the CEO, asked.

Householder waved a hand at Stenhouse and Oakes. "No more than they should have. I've been telling them for months that the whole idea is impractical. EU must have figured that out long ago."

Alvarez shot a hard look at Stenhouse. "Is that true, Professor?"

Stenhouse quailed before Alvarez's stare and hesitated. *This isn't my fault.* He squared his shoulders and sent his own angry glare at Householder. "I thought you and your so-called engineers could do the job. Mr. Oakes claimed you were competent."

Householder turned to Alvarez, his face red. "This is what happens when you let a scientist make engineering decisions. I told him it wouldn't work."

Alvarez turned to Oakes. "Did you know about this?"

"I knew Sam wasn't happy, but he assured me it would work." Oakes pointed at Stenhouse. "I thought Sam was griping because he wasn't lead engineer."

Alvarez nodded. "Well, starting now, he is the lead engineer." She smiled at Householder. "Will the Professor be useful as a consultant?"

"Are we going to continue the project?" Householder asked.

"Evaluating that will be your first task. I want a report by tomorrow afternoon on our course of action."

Householder glanced at Stenhouse. "I think Professor Stenhouse will be useful for a little while longer."

60

A WARM MARCH BREEZE and the resulting pleasant temperature convinced Correia to accept Senator Valerie Peterson's suggestion to have lunch at a restaurant with outside tables. Traffic was heavy on the adjoining street, but most vehicles were quiet enough not to inhibit conversation.

"Your poll numbers slipped a bit this week," Peterson said as they waited to give their lunch order.

Correia nodded and frowned as he looked at the people walking by. Many of them wore masks despite the breeze that had temporarily swept most of the usual pollution away. He and his companion kept their masks in their pockets.

With the election still more than a year and a half away, Correia's concern was for the party nomination first. His stance on making homosexuals contribute more to the country had gotten him support in the party, but others had jumped on his bandwagon. Some of them had been senators longer and had influence he hadn't had time to build.

"I need more than the gay issue," he said. "I had planned to use my support of General Power to show how friendly I was to business, but that won't do me much good now."

"You don't think they'll beat out this other company in the long run?"

Correia shrugged. "Who knows? The convention is more than a year away and anything could happen."

"Isn't Energy Unlimited in New York?"

"Yes, Schenectady. But if I try to exploit that, it could backfire and look like I'm favoring my constituents over the rest of the country. And according to their press release, the next two units they build are going to India."

"Nobody cares about favoritism. And production won't really start until they finish their factory this summer. You can talk about all the jobs that will create, and how much everyone will benefit from a new, clean energy source."

Correia stared at the foot traffic again. "Those people would like the 'clean' part. That won't help me get the nomination, but it could help in the election." He turned and looked at Peterson again. "It's an issue that will sound better to Priscilla, too."

"Emphasize job creation and the advantages of private enterprise over government subsidies. That won't hurt with the nomination but will look good in the election."

S TENHOUSE'S FIRST REACTION WAS anger. He was the discoverer of the Stenhouse Field, and they had relegated him to being a mere consultant. Hiding his resentment from his wife had been difficult, and he wasn't sure how well he had managed. Victoria was such a placid, accepting woman, reluctant to show any disagreement.

Mostly, he loved her for that, but he could have used someone to vent his frustration with. Instead, he held his complaints internally. Perhaps that delayed his eventual realization that blaming engineers like Householder may not have been fair. On the third day after the meeting, he walked through the engineering spaces on his way home and observed the team obviously concentrating on their jobs, trying their best. What experience did he have to justify judging them?

On his return home, Victoria examined him carefully after the usual hello kiss. "Your mood has changed." She wrinkled her nose and stared at him. "I'm not sure that it's for the better, though."

Stenhouse attempted to smile. "I was hoping you hadn't noticed. I've been trying to hide my anger at being demoted."

"So you're not angry now? It still bothers you."

"I've realized that I had no reason to be angry. That's what bothers me."

Victoria pushed him gently to a couch and sat down next to him. "So tell me about it."

"I'm a scientist. I shouldn't have been trying to manage an engineering project. The engineers tried to tell me what we were doing wrong, but I didn't listen."

"They still need you. You're the expert in the field."

Stenhouse shook his head. "Maybe. But I don't know how a girl just out of college with fewer resources did so much better."

"Don't you have a friend there? Maybe you should talk to him."

"McGavock? He's not a friend." Stenhouse paused and rubbed his chin. "Still, you're right. I should call him." He looked at his watch. "It's one hour later there. Probably a good time to catch him at home before he goes to bed."

"I can hold dinner off for a while."

Stenhouse settled in the comfort of his study before telling his phone to call Professor McGavock.

"Arthur, it's been a while," McGavock said when he answered. "I assume you're calling because of Energy Unlimited's announcement."

"It was quite a shock," Stenhouse answered. "I didn't think they could beat us to a product."

"Yes, I'm sure it was. So, is your position secure?"

Stenhouse shrugged. "I'm not the engineering manager anymore, but in retrospect, I think that's a good thing. I'm still employed as a consultant, which is probably more appropriate."

"Good. I'm sure that still pays better than being a college professor."

"I just don't understand how that girl did what I couldn't do."

Stenhouse heard McGavock chuckle. "I don't think she did, Arthur."

"What do you mean?"

"One of our mechanical engineering professors retired and went to work for them. I think he's the engineering manager, not Camila. At the least, he's a consultant and they've been paying attention."

"Which is what I should have done."

They talked for a few more minutes, but Stenhouse got off the call as soon as he could politely do it. He sat quietly, staring into space.

Life wasn't fair. Stenhouse had reached that conclusion many years before. His discovery of the Stenhouse Field should have brought him the success that accompanied other ground-breaking discoveries, but that hadn't happened. For years, most scientists ignored it or tried to discredit it, not accepting his claims. He had needed Ben Pinto's work in a weightless environment to prove he was right.

But it was Camila Lopez who made an obscure scientific discovery into what was a household word, or soon would be. By finding an important application for the field, she had forced its acceptance. And yet, no accolades went to him, the original researcher. He had hoped to change that with General Power, but had failed. While he still had his position as consultant, in reality, they had pushed him to the side.

He was sixty-six years old. He had sufficient savings to get by if he retired. Victoria would accept that, but he knew he couldn't. *My name should be up there with Einstein and Hawking.* Marking time with General Power wouldn't achieve that.

He saw only one chance to win the fame and fortune that was due to him. He had to swallow his pride and try to attach himself to Energy Unlimited. The irony of needing to humble himself to gain honor didn't escape him, but at his age, what else could he do?

61

R YAN SCROLLED THROUGH THE report from Martina about parts for the two generator units ordered by Basu Power. Equipment problems at the machine shop had slowed fabrication, but the parts were now ready. Marlee Dubois had already reported the completion of the control circuits and other electronic components, and final assembly could begin. It was Thursday, however, and Ryan suspected there would be people working over the weekend to meet the first delivery to Basu.

His secretary buzzed him. "You have a call from Arthur Stenhouse," he said.

That was unexpected, but Ryan told his secretary to put the call through.

"What can I do for you, Doctor?"

"First, let me congratulate you and your company," Stenhouse said. "Your achievement is quite impressive."

"Thank you. We're proud of our success in harnessing your discovery."

"Yes, my discovery. It was only a footnote in science until your people took it on. I owe a debt to both Ben and Camila."

Ryan could see where the conversation was going. *Have things gotten that bad at General Power?*

Stenhouse continued when Ryan didn't immediately answer. "I'm sure you're busy, Mr. Terry, so let me get to the point of this call. Energy Unlimited has become the leader in putting my work to use. I think I could be helpful, and would like to be considered for a position in your Research Department."

We have a Research Department? Not officially, although that was what Camila and Ben had become in the prior months. Making that official might be a good idea, but giving Stenhouse a part in it needed some thought.

"I think that's possible," Ryan said. "I'll be meeting the team tomorrow night, and I'll bring it up with them then."

"I appreciate that."

C AMILA LISTENED WITH MIXED feelings as Ryan told the group about his conversation with Doctor Stenhouse. It was encouraging to think that the discoverer of the Stenhouse Field would want to work with them. Her previous encounters with the scientist didn't inspire trust, however.

"Yesterday was April Fool's Day," Mike said when Ryan had finished presenting the request. "Are you sure he wasn't pranking you?"

Ryan looked at his son and smiled. "He doesn't strike me as a prankster. And he can't be happy about how things worked out for him at General Power."

"Did they fire him?" Camila asked.

"He didn't say that, and I didn't ask. I suspect they didn't, though. That would emphasize their failure."

"Stenhouse was engineering manager," Dean Curtis said. "Naming a scientist to manage engineers isn't a good idea."

"That's why we put up with you," Ben said.

Dean laughed. "And you're lucky to have me to put up with."

Camila shook her head and smiled. She remembered being intimidated by Dean when he was evaluating her thesis proposal. He had seemed to be such a gruff, humorless man. Had that been an act, or did he find working with them that much more rewarding than teaching? He certainly hadn't been gruff or humorless since joining them.

He was sitting next to Mary Welch, and she realized that was the norm for their group meetings. Mary was smiling at him, and Camila wondered if that might be another reason for the mechanical engineer's improved mood. He was decades older than the machinist, but Mary wasn't the same depressed woman either.

"So what are we going to do about it?" Ryan asked. "Even if we don't hire him, we probably should formalize a Research Division. Camila and Ben are already a de facto group."

Camila pulled away from her musings about Dean and Mary. "Are we? You all know about my interest in space flight. Since I can't help that much with the engineering work, I'm pursuing that. I think Ben's motives are similar."

Ben nodded. "Does that justify creating a formal group?"

"Maybe," Dean answered. "We're concentrating on small generators right now because that's the low-hanging fruit, but I think we'll move beyond that. Even the basic design could be adapted for large vessels. Ships, trains, even spacecraft. We'll always need research."

"Who would be in charge?" Dani asked. "If we hired Stenhouse, he would want to run it."

"That's not going to happen," Ryan answered. "I don't think we could trust him for that—at least, not until he proves himself." He looked at Camila.

Camila raised her hands. "Not me. Ben before me."

Ryan jumped in quickly, cutting off a comment from Ben. "You're right. If we make a Research Department part of the organization, Ben would be the best choice to lead it."

Ben frowned. "Two or even three people aren't a department. A Research Group."

Ryan nodded. "Okay, Research Group it is. And I take it you're agreeing to lead it." He looked around the table. "Any objections?"

Ben appeared to think about Ryan's statement. "Fine," he said after several seconds. "But what are we going to do about Stenhouse?"

"Being able to announce that he is joining us would be a publicity boost," Dani said.

"If that's all, he probably won't be happy and will move on," Ben countered. He turned to Camila. "What do you think? Would he be helpful in developing future applications?"

Camila hesitated. If they brought Stenhouse into the company, would he overshadow her? *But that's not what's important.* Would he help in her ultimate goal of developing space applications? She trusted Ben to keep their efforts moving in that direction.

She nodded. "I think so."

"He could be sent here to spy on us," Martina said.

"Our lawyers can deal with that," Ryan answered. "First, he'll be working with Ben and Camila on research, and we'll keep him away from our power generation technology, at least until we trust him. And we'll wrap him up tight with an iron-clad NDA."

Camila wasn't convinced that was enough. But she kept that to herself, knowing that Ryan was more experienced in such things than she was.

62

WHEN STENHOUSE JOINED GENERAL Power three years before, he moved from Spokane to Chicago where their research laboratories were located. Victoria Stenhouse didn't care for the more urban environment compared to eastern Washington, but she looked forward to living in upstate New York and that helped him adjust to his new work situation.

Meanwhile, he had to find his place at Energy Unlimited amid everyone already part of what appeared to be a close-knit group. Ben Pinto hadn't made a big deal of it during interviews, but left no doubt that he would be in charge, not Stenhouse. Any resentment Stenhouse felt disappeared as he realized how much Energy Unlimited's success came about because his two new colleagues had so completely given up leadership to Dean Curtis for what became an engineering and manufacturing company. That only confirmed what he believed about his failure at General Power.

He would have to win their acceptance. The restrictive NDA he had to sign before being hired left no doubt about that, and despite the decision to hire him, it was clear Camila Lopez and Ben Pinto didn't trust him yet.

Dani Terry, Ryan's daughter-in-law, created a splashy publicity campaign around his move to Energy Unlimited. For now, that was his value to the company, presenting them as a new major player in the energy industry. It would take time to show that he could contribute more.

Ben scheduled a meeting for his first day, and Stenhouse looked forward to the chance to make an impression on the two younger scientists. He had thought about little else since accepting their offer of employment two weeks before, and even Victoria had noticed his new enthusiasm.

"Ryan called a board meeting for next week," Ben announced to begin the meeting. "Amrita Savant will come to represent Basu's interests, and I want to present our research goals going forward.

"Basu Power now has two of our generators, and they're enthusiastic about their future. I expect them to be asking for more. The completion of our factory next month will be an important subject at the board meeting, but there will be questions about our research too."

"Does your relationship with Basu Power present any problems in setting our goals?" Stenhouse asked.

"Not technically. Jürgen and Ryan may have a contrary opinion," Ben replied.

Camila laughed. "Jürgen relies on me to keep him safe."

Stenhouse didn't understand the comments, and assumed they were part of some inside joke. He smiled because he thought that was an appropriate response.

"We've been vague about what we're doing until now," Ben continued. "Mostly, Cam and I have been taking classes to broaden our knowledge. We should get more detailed about this group's direction."

"I guess dreaming of spaceships isn't realistic," Camila said.

Ben nodded. "Not yet. I have a suggestion for a research topic that might sound good to Basu and maybe outside the company. We're using lead as a fuel because it's cheap and more efficient than most other possibilities. But there are other ways to think about it. We could use radioactive elements as fuel, too."

"What's the advantage of that?" Stenhouse asked. "Uranium and other fuel sources are more expensive and require special handling."

"Material suitable for fission reactors is expensive. But they produce waste products heavier than lead. A better disposal method, like converting the waste to harmless iron, would be welcome."

Stenhouse wasn't sure. "Could your generators use enough to make a difference? You measure fuel consumption in pounds."

Ben shrugged. "Maybe not, but that's where our research comes in. It would be our task to find out."

"Existing fission power plants could install them, solving their waste disposal problems and generating more power at the same time," Camila said.

"Exactly," Ben said.

"I would like to investigate using the end products for propulsion as well," Camila added.

Stenhouse didn't immediately understand. "Propulsion?"

"Spacecraft use ion drives that have a high specific impulse because they expel particles at higher speeds than chemical rockets," Ben explained. "But the low mass means low acceleration, so it takes a long time to reach the velocity they are capable of."

"A propulsion system that expels heavy iron ions at the same high velocity would be more effective," Camila added. "Ben and I have talked about this before. My original goal was to use your discovery for space travel."

Stenhouse nodded. "I see. Those are two very different directions for our research. It should be interesting."

B EN PRESENTED THEIR PLANS at the board meeting. He spent more time on the proposal to design a system for nuclear fission waste disposal because that had more immediate application. He wasn't surprised when Amrita Savant focused on that when he finished.

"The emphasis for the last ten years has been fusion power." She waved her hands in emphasis. "This could change that. Fission is a more mature technology and less expensive, but is often rejected because of the waste problem."

"Ultimately, we see Stenhouse power replacing both nuclear technologies," Ryan said. "But Ben's proposal could be an important interim path to clean power generation."

Amrita nodded. "One reason I came was to discuss building a large generation facility in India. We need it badly. But that will take years to plan and build." She paused and looked around the table. "I hope you will pursue this idea vigorously."

During the ensuing discussion, the questions involved timetables and resource application. Everyone agreed the proposal had tremendous potential, and the vote eventually taken hardly seemed necessary.

None of that deterred Amrita from expressing Basu Power's desire to begin work on a significant generating facility in India. That was more an engineering issue, and Dean promised a preliminary evaluation of a project before the next board meeting.

63

I T WASN'T NECESSARY TO check the weather forecast to know the day would be hot. The July factory opening was scheduled to begin at nine in the morning to avoid the worst of the heat. Still, the temperature was already in the eighties. Pollution levels were high enough to make masks advisable, adding to the discomfort of the attendees.

Later, a light lunch would be served inside where there would be air conditioning, but the requisite speeches were planned for the outside, where a raised stand had been set up for the speakers.

Senator Correia joined other dignitaries on the stand and tried to look attentive as Ryan Terry spoke about Energy Unlimited and asked each of its major officers to stand when he introduced them. At least he kept his presentation short.

Terry gave a brief introduction of Doctor Stenhouse before motioning the lovely woman next to him to rise. He described Camila Lopez as the inspiration for the company, but didn't mention a title. *She's quite beautiful.* Biographical information made available to attendees said that Lopez was single. She was in her mid-twenties, and Correia found it mildly curious that she wasn't married.

After Terry, the governor of New York made a longer speech that strained Correia's ability to feign interest. Since the senior New York senator hadn't come, he would speak next and had to stay alert.

By the time the governor was done, it was after ten and the temperature had risen noticeably. There was no breeze, but when Correia removed his mask for his speech, the burned smell in the air from nearby facilities wrinkled his nose. At least the factory they were celebrating would not add to the pollution, but it was contrary to his party's talking points to mention that in his speech.

The district's congressman and the mayor of Schenectady would follow him, and he had decided to have mercy on the crowd and keep his speech short. Rather than talk about the environmental benefits of the Stenhouse generator, he reiterated the governor's points about area employment, implying that he had supported the project since he first learned about it. There was some truth to that, although until recently, he had conferred his attentions on General Power, not Energy Unlimited.

It was almost noon before the speeches were over, and they moved into the factory. Chairs provided seating on a large open area of the factory floor, with food laid out on tables. In the cool confines of the building, people relaxed, got food, and formed small groups. Correia wandered for a few minutes, greeting and shaking hands, but stopped when he saw Camila Lopez sitting with a man he didn't recognize.

He joined them, ignoring the frown on her companion's face. "Miss Lopez, I'm Senator Correia."

Camila stood and shook hands when he held his out. "Hello, Senator. I'm so happy you could join us."

"It's my pleasure."

The man with Camila also rose, and Correia saw his frown become an unconvincing smile.

Correia was charming when he wanted to be, and he made sure the handshake with the man was warm and friendly. "Are you also part of Energy Unlimited?"

"Jürgen Varela. No, just an investor. My sister is Gabrielle Pinto."

"Ah. The software engineer."

"Yes."

When Correia saw the two sitting by themselves, he had assumed they were a couple, perhaps satisfying his curiosity about Camila Lopez's marital status. Now he wasn't sure. They seemed friendly enough, and Varela seemed to resent the intrusion. But the vibe wasn't quite right for two people romantically involved.

Perhaps Lopez was merely being friendly with one of the company's investors. The Indian power company was their biggest investor, but they could have others. Apparently, Jürgen Varela was one of them.

After a little more small talk, he excused himself and wandered away. Jürgen Varela had not warmed to him, and he sensed reticence behind Camila Lopez's smile. He wondered why.

G REG MORA GRINNED AT his roommate, and Bob Gamble grinned back. A dozen other men stood with them outside the bar where Pride Alliance was holding their August meeting. Some had signs, but mostly they made noise and harassed people entering or leaving.

The front of the bar looked like a place where homosexuals might hang out. A colorful awning shaded the entrance: not rainbow, but a bright red. Tall flowering plant in stone pots about two feet high flanked the door. *Who puts flowers in front of a bar?* A girl he once dated had told him she thought they looked nice. He wasn't dating her anymore.

"Having fun yet?" Gamble asked.

"You bet." Greg paused and watched an angry-looking man come out of the bar. "Uh oh, I'll bet that's the owner. He must be sorry he let those fags use his back room."

"All right, enough is enough!" the newcomer shouted. "If you idiots don't disperse, I'm calling the cops."

Gamble laughed and approached the man. "We're just exercising our First Amendment right to free speech on a public street." He moved forward until the two men were almost touching. "The cops should be called for your customers, not us."

"You tell him, Bob," Greg said.

The man turned red and glared down at Gamble. He was a big man, at least six inches taller than Gamble, but Gamble wasn't intimidated.

"If you want us to leave, you're going to have to make us." Gamble raised a hand to the other's chest and pushed. "And stand back. I can smell your breath."

Instead, the other man growled an expletive and pushed back. Gamble stumbled and tried to keep his footing, but he backed into a planter and lost his balance. Greg reached out to steady him, but missed, and Gamble fell, hitting his head against the planter with a loud crack.

Silence fell over the crowd, but it only lasted a few seconds. Gamble didn't move, and blood already stained the sidewalk. Greg stared at his friend, unsure what to do, but he heard sirens coming closer and backed away.

Several more people stopped to see what had happened, and he joined them, trying not to be noticed as a police car parked in front of the bar. He watched with growing anger as the two policemen examined Gamble.

"He's gone," the older policeman said.

Greg clenched his teeth and tightened his hands into fists. He couldn't do anything now, but he would have his chance.

F IVE WEEKS PASSED SINCE the factory opening, and Senator Correia was home again. After a delicious dinner, an evening walk through his upscale Virginia neighborhood, maybe even talking Priscilla into joining him, could have been enjoyable. But Virginia in August was even hotter than New York in July had been. Any thought of a pleasantly cool twilight stroll was a fantasy.

His presidential prospects were looking up. There had been some grumbling from the party after his appearance at the Energy Unlimited event, but that subsided when moderate commentators suggested his overall favorability rating had improved. So, he sat comfortably in his home office and turned on the news.

A story about the latest saber-rattling from the Russian Federation was business as usual. He listened, but pleasant memories of Priscilla's cooking distracted him. The anchor's announcement of another anti-gay demonstration in Troy brought him back to the news, and he leaned forward.

"We're getting reports now about violence outside a meeting of gay activists in Troy, New York," the anchor opened. "We know demonstrators assembled outside the meeting place and a fight broke out when someone from the meeting asked them to leave. There are reports of injuries, and we are trying to get details.

"This is only the latest in these disruptions and is reminiscent of a similar incident in Troy a year and a half ago when several people were injured. Similar confrontations have become more common throughout the United States."

"Is that another report of violence?" Priscilla shouted from the kitchen.

"Troy again," he responded. "Nothing to worry about."

But it might be something he could use in his campaign for his party's presidential nomination. When the report finished, he turned off the television and went to his desk to write a few notes about how he would do that.

64

D IEGO LOPEZ HAD A lot to think about when his sister came home for the Christmas holiday and brought Jürgen Varela with her. He couldn't avoid some resentment about Jürgen. After he sold his company almost four years ago, the new owners hadn't been interested in a software tester living in Texas. Working with V-World for six months had been fun, but after Jürgen abandoned his company, Diego had to settle for more mundane employment as an electrician.

His sister's relationship with Jürgen confused him, too. Camila and Jürgen were living together, and his parents were surprisingly unconcerned about that. He was almost twenty-six years old and lived in his own apartment, but they left no doubt about their reservations about his dating activities.

But Camila slept in her old bedroom and Jürgen slept in Diego's former bedroom. Camila told him they were housemates with separate rooms, a claim Diego hadn't believed at first, but maybe it was true.

Jürgen sold his company to be the first major investor in Energy Unlimited, before EU even had a product. It seemed incredible Jürgen would do that for Camila if they were just friends.

Of course, from what Camila said about Energy Unlimited, the investment had been a smart move. Their first product was selling well. His sister was enthusiastic too about her research on a larger model that could convert tons of dangerous nuclear waste into electricity and nontoxic iron. The benefits of that were obvious.

Energy Unlimited dominated the conversation for the first day after Camila's arrival. Some of her explanations were technical and, while Diego understood most of it, he didn't think his parents did. Tomás and Nina paid close attention, but their occasional nods and polite sounds didn't fool him.

The next day was Christmas, celebrated with his mother's typical elaborate dinner spread of the family's favorite Hispanic dishes. The meal began in the early afternoon, but the sun was low in the sky by the time they left the table. Everyone helped with the cleanup before sinking into the living room furniture to contemplate their overstuffed condition.

Diego was not alone in feeling drowsy, but his father wanted to talk.

"Supposedly, your Senator Correia wants to run for president," Tomás said.

Jürgen nodded. "I think so. It's interesting how he's using Energy Unlimited, talking about the product Camila is researching for nuclear waste disposal. He can't admit we still have pollution problems, so he touts it as an example of how free enterprise has shown that the government doesn't need to be involved."

"And of course, making sure that everyone knows your company is in his state," Tomás said.

"Of course." Jürgen grinned.

Tomás shook his head. "That's all fine, but I won't vote for the man."

"I certainly won't either," Nina said.

"The way he's tried to capitalize on anti-gay attitudes is despicable," Tomás said. "He just encourages violence with his rhetoric. Like that incident in August where you live. Implying homosexuals killed a protester in retaliation. He makes it sound like that bigot was a victim."

"The death was called an accident," Camila said. "The man who pushed him wasn't even gay. He was the owner of the bar, asking them to stop harassing his customers."

Tomás grimaced. "That doesn't get mentioned often. And certainly not by Correia."

"He doesn't talk about an earlier attack in Troy either," Jürgen said. "Several people, all gays, were injured and someone Camila knew was hospitalized."

Diego saw his sister frown and shake her head. *A bad memory?* "Is she okay now?" he asked.

"She recovered," Camila answered. "Maybe we could change the subject."

S ENATOR CORREIA SMILED AS his favorite supporter approached. Fellow Senator Valerie Peterson smiled back.

"You look happy," she said. "Good news?"

"Mostly," Correia answered. "I spoke to the party head this morning, and I think that, after my victory in New Hampshire, he'll support me."

"He's not worried about how close it was?"

"He is, but a win is a win, and he thinks resurrecting that bill we tried to pass in 2051 will help with the South Carolina primary. If a New Yorker like me can at least come close there, it will boost my chances."

"You're talking about the bill to tax homosexuals because they don't have children."

"That's not exactly how I describe it. It doesn't apply to homosexuals only, but yes. That bill. He thinks it will resonate with a lot of voters." Correia paused. "He was less happy about my support for nuclear power. According to him, it sounds like I admit there's still a climate problem, regardless of how much I talk about jobs and the power of free enterprise."

Peterson nodded. "That will play better to the general electorate than to the party."

"I told him that. He suggested I downplay it until I have the nomination. He's probably right."

"I'll support you, but I don't think I can make it too obvious. A lot of my constituency still hasn't adjusted to oil losing its dominance in energy production."

"Three decades later. Yes, you Texans can be stubborn."

She shrugged. "You go with what you have. How's Priscilla taking all this?"

Correia scowled. "That's a problem. She hasn't been happy about my proposal to make homosexuals contribute."

"Doesn't the prospect of becoming First Lady help her get past that?"

"I don't think so. Last week, she said something about selling my soul."

"Hell, Leo, we did that a long time ago."

A PRIL BEGAN WITH TEMPERATURES in the eighties. Greg Mora watched a show about physical fitness urging viewers to get out and walk more. That was a good idea, but he would have to wear a mask and then he would sweat. That was too much trouble, so he changed to a news report and reached for the bag of potato chips on the couch next to him.

A report on Senator Correia's campaign described the upturn in his primary success and predicted he would be nominated for president. According to the reporter, the senator's speeches about forcing homosexuals to contribute their fair share to the country drove his rise.

Greg munched on a chip. "Damn right."

Bob's death was eight months in the past, and the newscast reminded him he had done nothing. He attended demonstrations occasionally, but that accomplished little. Mostly, he watched other people do things. He glared at the bag of potato chips, closed it, and threw it across the couch.

The subject shifted to Correia's support of current projects confirming the senator's stance in favor of solutions through free enterprise rather than government programs. It mentioned his endorsement of the work Energy Unlimited did to provide innovative solutions for the country's power needs.

Greg frowned at that. Energy Unlimited was Camila's company. It was Bob who had told him she knew the girl injured in that other demonstration, the first he had attended with Bob. He hadn't done anything about that either, although he wasn't sure what he could do.

It wouldn't help Camila's company if she were outed, but he needed more evidence. Second hand information from a dead man about a single encounter wasn't nearly enough. Was she still taking classes at the college?

He had the weekends and vacation days coming to him. It was difficult to get any kind of personal information about people anymore, but if he could learn some of her habits, maybe he could follow her and see who she associated with. He knew Energy Unlimited was headquartered in Schenectady, and he could start there.

"Sounds like a plan." Greg grinned and turned off the newscast.

65

R YAN LIFTED HIS GLASS of beer. "Here's to our success. Congratulations to all of you."

Around the conference table, a dozen other glasses were raised. The usual Friday gathering had started with a glowing report from Mike Terry about sales of their original generator design. Then Dean Curtis announced manufacturing would begin within the month on a larger unit meant for processing nuclear power plant waste. Mike added that they already had half a dozen orders for that product.

Less spectacular, but still welcome, Mary Welch reported her machine shop was accepting more business now that the company's factory was producing, and parts for Stenhouse generators no longer absorbed the shop's capacity.

Ryan lowered his glass. "To honor all this good news, I've arranged something a little fancier for next week's meeting. I've made reservations for all of us at La Parisienne. Sahil Nanda will also come from India."

Several people gasped. La Parisienne was probably the most expensive restaurant in Albany, and Ryan hadn't been there himself. Jürgen had recommended it, having taken Camila there twice.

"I'm also going to suggest to the Board that the employees who can't celebrate with us next week get substantial bonuses this year," Ryan continued. "Dani is working on a plan for that to be presented at a board meeting next Thursday."

W HEN HIS PHONE ANNOUNCED the call, Ben didn't immediately pay attention. He didn't recognize the caller's name, and the problem on his workstation held his attention. He assumed, without really thinking about it, that the caller would leave a message if the subject was important. Five minutes later, Stenhouse entered the office.

"They called you?" he asked.

The scientist's voice was unusually loud, but not from anger. Ben looked away from the screen and Stenhouse was grinning and waving his arms.

He looked at Ben and shook his head. "No, you couldn't have heard. You must have a message, though. Check your phone."

"Okay." He listened to his phone, but didn't believe the message at first. Then he was standing and matching Stenhouse smile for smile. When the scientist threw his arms around Ben, Ben returned the hug.

SOMETHING WAS GOING ON. Camila had expected Ben and Gabby to share a taxi with Jürgen and her on the ride from the EU office to the restaurant, but it was Gabby and Martina that joined them in the vehicle. Ben was in another taxi with Ryan and Stenhouse.

Camila turned to Gabby. "What's going on? Why isn't Ben with you?"

Gabby shook her head. "Darned if I know. He's been weird for two days now, and he's suddenly best friends with Doctor Stenhouse. He won't tell me why."

"I think Ryan knows," Martina said. "All three of them look like Christmas came early."

"Ben told me everyone would find out at the restaurant," Gabby said.

Even the seating at the restaurant was unusual. The host ushered them into a private room already set up for them and Ryan took the head of the table with Ben and Stenhouse on either side. Gabby sat next to her husband, but he was paying more attention to the quiet conversation Ryan and Stenhouse were holding. Professor Moreno sat next to Stenhouse, looking a little awestruck, but she was also ignored.

The host had not distributed menus, but he left, closing a door behind him. Ryan stood and glanced at the door before turning to the group.

"Last week, I stood before many of you and announced this celebration. We had much to celebrate then, but two days ago, I learned that we have one more reason. Before I tell you, I want to remind you we're in a very classy place and some of you," and he looked at Mike and Dani, "should try to restrain your usual exuberance."

Ben caught Camila's eye, smiled, and seemed to mouth, "I'm sorry." That only increased her bafflement. But Ryan continued.

"In retrospect, this development is only logical. We really shouldn't be surprised." He frowned and shrugged. "I was, though, so maybe you will be as well."

"So get on with it, dad," Mike growled.

"Don't rush me," Ryan scolded. He paused. "All right, I guess I'll just come out with it."

"Please," Mike said.

Ryan nodded. "Two days ago, Doctors Arthur Stenhouse and Benjamin Pinto were informed that they would receive this year's Nobel Prize in physics."

G REG DIDN'T HAVE MUCH luck at first. He took a day off from his job a few times, hired an automated taxi, staked out the office building for Energy Unlimited, and observed Camila leaving. Most days, an automated taxi picked her up, and Greg couldn't tell his vehicle to follow her. Then one day, Jürgen Varela met her instead. He still couldn't follow them, but that was the break he needed.

The architectural firm Greg worked for had built Varela's house before Greg joined the company, and they still had records from that project. That gave him an address, and on an October evening, he played a hunch and parked near Varela's house. His suspicion was confirmed when he observed Camila arrive and admit herself. At the least, she was a frequent and welcome visitor, but was probably living with Jürgen.

That made Greg doubt his belief that Camila was gay, but only briefly. He knew Varela was the sister of one of Camila's friends and an investor in Camila's company. Living with Jürgen was a matter of convenience, not romantic attachment. After all, she had rejected him before she met Jürgen. And he had Bob's sighting of Camila with another lesbian.

After five months of learning nothing, he finally had some useful information. Using the automated taxis limited him, since he had to give them a destination before they would take him anywhere. To get more, he would have to follow Camila, so he rented his own car for a Friday night so that he could control the vehicle's movements. Then he parked outside Camila's office building.

It was one of those nights that Varela picked her up, but two other people also got into the car. When their car moved away, it didn't drive toward Varela's home. Greg smiled and congratulated himself on another hunch that paid off. He followed them.

When their car got on the Thruway heading southeast, he guessed they were going to Albany, and again he was right. They stopped at a restaurant on a busy downtown block, and Greg could see them meet several others at the entrance. Everyone looked happy, apparently celebrating.

He wanted to wait, even though it would be a couple of hours before they came out again. Parking was difficult, but he had time, and the car found a space with a view of the entrance after half an hour of driving around the block. Then he used his phone to look up the restaurant. As he expected, La Parisienne looked and sounded expensive, discouraging him from going inside.

Games on his phone amused him for a while, but after the first hour, he was bored. *I should just go home.* But he didn't. When several taxis pulled up and a group of people exited the restaurant, he sat up and stared out at them. Yes, there were Camila and Varela, talking to another man. He recognized the girl who had tripped him years ago, and some others also looked familiar.

Without thinking about it, he got out of his car and walked toward the group. No one noticed him until he was about ten feet away. Then Camila looked at him, and he could see the startled look on her face. *At least she still recognizes me.*

"WHY DID YOU APOLOGIZE?" Camila asked Ben as they left the restaurant.

"You should have been included," Ben answered. "Your work is just as important."

Camila shook her head. "It was Doctor Stenhouse who made the discovery and your research that convinced everyone he was right. The Nobel Committee made the right decision."

Gabby took Ben's arm and pushed herself close. "And I'm so proud of my hubby."

"You should be." Camila sensed someone approaching and turned. "Greg," she said. She frowned.

GREG SMILED. "HI, CAMILA. Long time, no see."

Varela looked at him and frowned. Then he took Camila's arm and urged her toward a taxi.

"Running off so soon. It's been so long since. . ." He stopped as another man in the group stepped between them.

"Not long enough, apparently," the man said.

Greg looked at the man. He could try to push him out of the way, but he was bigger and the attempt, successful or not, would make a scene. He backed away a couple of feet.

"I just wanted to say hello." But Camila was already in the car. Varela turned and glared at him before he joined her.

Shaking his head, Greg turned and walked away. "They're probably all gay," he muttered.

66

A FTER THE DINNER AT La Parisienne, normalcy returned for several weeks. But it was an election year, and November demanded attention. Camila watched the early election results on a screen in the conference room. Polls weren't closed in New York, and the West Coast would be even later, but the early returns left little doubt that Leo Correia would be the next President of the United States.

Ben, Gabi, and Marlee Dubois worked in the EU building and Ryan, Mike, Dani, and Jürgen arrived late in the afternoon. Correia's support of Energy Unlimited in his campaign had probably helped the company's success, and most of her assembled friends were celebrating. Jürgen was solemn, however, and Dani's and Gabi's reactions were muted compared to the others.

The usual time for going home passed, but everyone remained, waiting for the inevitable victory speech. It came shortly after seven, a typical litany of self-congratulation and promise. But one line stood out.

Correia stared out at a crowd of supporters, looking both jubilant and firm. "This election has proved that this country wants someone who will support free enterprise and guarantee that everyone contributes their fair share to its success."

To most people, that statement would be as innocuous as the rest of the speech. Camila glanced at Jürgen. His frown confirmed that he too was thinking beyond the bland statement to its chilling implications.

A week later, the election was old news, and Correia was inaugurated two months later. For Energy Unlimited's principal members, another event seized their attention; Stockholm beckoned.

O N THE MORNING OF the ceremony, held on Alfred Nobel's birthday, Ben thought the worst was over. Two days before, he had stood before an audience of royalty, fellow scientists, and other influential people presenting his Nobel Prize lecture, following Doctor Stenhouse's address with a talk about his orbital experiments. He was receiving the prize for confirming Stenhouse's work, not for progress in understanding dark matter, and his speech touched only lightly on that.

For someone who had never talked to a group that large or that prominent, the experience was stressful. The rest of the EU Board was in the audience, and they assured him afterward that he had done well, but he was only relieved it was over.

Sitting on the stage of the Stockholm Concert Hall, waiting for the prize ceremony to start, he felt just as nervous. As a full orchestra played Mozart on the balcony behind him, he stared at the crowd of a thousand eminent personages seated on the hall floor below the stage or on the balcony above. Gabby and the others were out there somewhere, but he couldn't spot them. He wanted to turn and see if they were in the seats behind him, but could only sit uncomfortably, trying to let the music calm him.

The King of Sweden and the royal family walked onto the stage and took their seats. The ceremony began officially with a brief speech from the President of the Royal Swedish Academy of Sciences, Doctor Hans Olsson.

"Your Majesty, your royal highnesses, esteemed Nobel Prize laureates, ladies, and gentlemen," Olsson began. "It is my pleasure to welcome you on behalf of the Nobel Foundation. The laureates we honor today are the hope of our civilization, advancing our knowledge to bring hope of solving problems that have afflicted us for all of history. Such was the intention of Alfred Nobel when he created this award."

Olsson continued with a brief acknowledgment of the Peace Prize ceremony that had taken place previously. That award had been given to two politicians, one Israeli and one Gazan, for their efforts in finally establishing a stable Palestinian state. Olsson spoke for several minutes about the hope engendered by their work.

"But political and religious differences are not the only source of conflict in our world," Olsson continued. "The struggle for resources has existed for humanity's entire history, possibly before politics or religion. The awards presented today are notable for the hope they give for amelioration of the scarcities that have motivated so many wars. We look forward to the lives improved by the work honored and pray that, unlike other advances, the result will be primarily to the benefit of humanity."

Olsson sat down and the orchestra began playing again. Ben had never been a fan of classical music, but hearing it now, played by world-class musicians in a setting optimized for such a performance, he relaxed just a little. He glanced at Stenhouse sitting next to him, but his eyes were closed, probably enjoying the music as much as Ben.

The music ended and the presenter of the physics award, Doctor Ingrid Ericson, came to the podium. After addressing the audience, she continued. "Major advances in the sciences are celebrated, but can take decades before revealing their practical benefits. The discovery of the Stenhouse Field by Doctor Arthur Stenhouse fifteen years ago was such an advance. The detection of the previously unknown phenomenon and the means of controlling it was so incredible that many did not accept it at first. The effect was subtle, almost impossible to duplicate, and was largely ignored.

"It was seven years before a young graduate student, Benjamin Pinto, had enough faith in the discovery to make it his thesis project. Doctor Pinto took a Stenhouse generator into orbit where, free of the disruption of Earth's gravitation field, he proved the field did exist.

"That alone would justify the awarding of the Nobel Prize for physics. But they haven't stopped there. Doctor Olsson spoke about conflicts caused by resource scarcity. Doctors Stenhouse and Pinto now work in the Research Group of a company dedicated to harnessing the Stenhouse field to produce inexhaustible energy from the commonest of materials."

Ericson paused and looked at Ben and Stenhouse. "The King of Sweden will now present the Nobel Prize in physics to Doctor Arthur Stenhouse and Doctor Benjamin Pinto."

Stenhouse didn't look nervous when they stood. Everyone else on the stage also rose, and the king stepped to the middle of the stage. Stenhouse smiled and approached the king, looking confident that he belonged there. He accepted the black box containing the medal and shook hands with the king. Still smiling, he backed away, bowed to the king, and then bowed in each direction to the audience before returning to his seat.

It was Ben's turn. As the president announced him, he stood and walked slowly toward the king. *Stand straight!* The king smiled at him as if he could see Ben's anxiety. When he held out the medal, Ben stared at it for several seconds, trying to understand how the gold medallion embossed with an image of Alfred Nobel could be for him. Afterward, he was sure he had thanked the king, but he couldn't remember it. He only vaguely remembered his own bows before he sat down again.

The physics prizes were awarded first. Ben sat stiffly, occasionally looking at the award again, as the other prizes were presented, ending with the prize in Economics

Sciences. He breathed a little easier as the orchestra played the Swedish National Anthem, and the royal family left the stage. As the orchestra continued playing, other notables on the stage filed by, shaking hands with the laureates, but now the worst really was over.

Of course, there was still the formal banquet that evening.

67

R YAN CALLED THE BOARD to order. "I scheduled this meeting for today because it's the one-year anniversary of the last time we were all together, the opening of our factory here in Schenectady. I'm sure you'll all agree that it has been an incredible year. Later, Danielle Terry will join us to present the latest sales figures for our Mark One unit and the Waste Conversion product. I can tell you now they are impressive.

"We have many other things to talk about, but there are two especially important items on the agenda. The first is related to the sales data we will see. Our existing factory is inadequate for the demand we expect."

He paused and looked at Sahil. "The second item is a proposal from our Indian partners to build a complete power facility in India, relying entirely on Stenhouse technology. Such a task will mean hiring many more people, including civil engineers."

Sahil interrupted. "We already have many such working for Basu Power."

Ryan nodded. "That may be the best way for this first facility. Eventually, if we are to expand in similar projects outside of India, we will probably need in-house engineering."

"Then you are open to a facility built by our workers," Sahil said.

"Well, we'll discuss that later. We should consider that seriously as a plan more likely to be successful. We'll hear from Dean Curtis when we talk about a second factory. He'll address the issues better, but I suspect we'll need more engineers as well as manufacturing capacity. We'll need new products with greater output, I would think."

"If you shared the designs of your products with Basu, we could help there as well."

Ryan smiled. "That too can be a topic for later conversation." *And I'm not looking forward to that.*

RYAN EXPLAINED THAT A second factory was a necessity if Energy Unlimited was going to grow. But the success that created the need was also the solution. Ryan already had investors lined up and was negotiating terms. The Board discussed what terms would be most advantageous, but there was no disagreement with the proposal. A facility in India was more controversial, requiring more discussion. After taking Sunday off, they reconvened on Monday.

Dean Curtis, Engineering Department Manager for the company, joined the meeting. "I agree it would be more expeditious for the civil engineering to be done by your people. It will require coordination with the people here who will design the generation modules, but an exchange of information will help us develop the abilities we'll need for future projects."

"Basu Power is a part owner of Energy Unlimited," Amrita said. "We should have access to the technical data that we help produce. If we had that, we could design the new generators as well."

Ryan shook his head. "Our government would resist that. I'm sorry, but the technology must be kept internally."

"That recalcitrance isn't what Basu paid half a billion dollars for," Sahil said.

"No, you paid it to get first dibs on the products we develop and a share of the profits." Ryan paused. "Let's see, in the first year, your profits have already amounted to ten percent of your investment. That will increase, and you should pass the breakeven point in less than five years. Of course, that's in addition to the benefits you will accrue from the use of the technology."

Amrita glared at Ryan. "So the bottom line is that you will gain knowledge from our engineers, but will contribute nothing to our knowledge."

"If you choose to think of it that way." Ryan shook his head. "It would be more accurate to say that the bottom line is the growth of your company compared to your competition because of your access to the technology. That's what you agreed to when we became partners."

Amrita clearly wanted to say more, but Sahil sighed and put a hand on her arm. "You have a point. Let's talk about how we will coordinate the effort."

Dean took over again. "Building a facility from scratch would be a massive effort, probably prohibitively so for Energy Unlimited right now. We've discussed a project before to process the waste products from India's existing fission facilities. We could easily double the output of those facilities and do it much earlier."

Amrita frowned. "Using Stenhouse reactors manufactured here."

"We already have the design for those units," Ryan said. "Tooling for their manufacture will be ready within a month."

Sahil nodded. "Energy Unlimited is still new and small for such an ambitious undertaking. Basu's leadership will have to accept that for now. So we would prepare a site to add waste-processing reactors you would manufacture."

"I think that would be the best way forward," Dean said.

68

WITH NO NEW REACTOR designs requiring research, the EU Research Group had no pressing work to do. That gave them the opportunity to consider future projects.

"Looking forward, I think we should start talking about space applications," Camila said.

Ben grinned. "Shocking."

"We've already talked about something similar to an ion drive," Doctor Stenhouse said. "That would be a suitable topic for research."

After two years, he was still Doctor Stenhouse to Camila. Ryan had addressed him as Arthur once in his first month, and the obvious disapproval made the scientist's opinion of the familiarity clear. If Ryan got that reaction, Camila wasn't brave enough to be less formal. She had heard Ben address him as "Art," but sharing a Nobel Prize had bonded the two men.

"Yes, we should start there," Camila said. "There's another issue we should discuss, as well." She paused. "Theoretically, a Stenhouse drive could accelerate a vessel to a significant fraction of the speed of light. That could create additional problems."

Ben nodded. "Friction."

"Exactly. Space, especially within the solar system, isn't a perfect vacuum. It's close enough that traveling a few thousand miles an hour isn't a problem, but I think it would be at millions of miles an hour."

"Erosion of the hull would limit a spaceship," Stenhouse said. "Still, velocities an order of magnitude better than existing spacecraft should be possible."

"That would be a big help for supporting a permanent presence on Mars," Ben suggested.

"In the near term." Camila leaned forward. "I want to think beyond that."

Stenhouse raised his eyebrows. "What?"

Camila hoped she looked confident. "I think we could use the Stenhouse Field to shield a ship, allowing it to approach light speed."

Ben chuckled. "You've been watching old Star Trek episodes again. 'Raise shields!'"

Stenhouse seemed to take her statement more seriously. "Such a field would only have to deflect molecules away from a ship, not laser blasts. But the ship would need an enormous field. Is that possible?"

Camila smiled. "That's what I want to research."

"We can do that," Ben said. "But I have another suggestion. Experiments with broadcasting focused microwave beams to ground-based receivers have shown that to be practical. I would like to investigate having orbiting reactors to generate those beams."

"Fueling them could be tricky," Camila said.

Ben nodded. "Especially if the fuel had to be brought up from Earth. But what about the moon? Meteors and asteroids perhaps."

Stenhouse shook his head. "Elements heavier than iron are not common in those bodies."

"True. But not nonexistent. Molecular sieves could separate the fissionable material from the vaporized fuel before it enters the reaction chamber."

Stenhouse looked thoughtful, but not convinced. "Perhaps."

Camila could feel the excitement building. "I guess we've got plenty to do."

"We can all work together, but there's three of us and three ideas," Ben said. "I'll take lead on the space reactors."

"I think I should investigate Camila's shield idea," Stenhouse said. "I've had the most experience in working with a Stenhouse Field."

Camila wasn't sure she agreed. She had gotten a lot of experience simulating field configurations working on her thesis, and she had more experience using Gabby's software in those calculations. But Ben said they would work together, implying she would at least contribute to work on a shield. Perhaps Stenhouse was right.

"Then a spaceship engine is my project," she said.

Ben smiled at her. "I suspect that would be an engineering project as much as a physics project. You have some experience with that."

"Initially, perhaps. Having an engineer in the group to keep us grounded might be a good idea soon," Camila said.

Stenhouse nodded. "Yes, I agree."

S INCE CONFRONTING CAMILA OUTSIDE the restaurant in Albany, Greg Mora had
tried to forget about her. For a while, he had been contented to cheer on Senator Leo
Correia in his campaign for president. After the inauguration, Greg expected changes that
would bring some manner of justice to people like Camila and Jürgen. In his obsession,
he didn't ask what form that justice would take or what it was that required justice.

When Correia's tax bill foundered, Greg's resentment increased. It didn't occur
to him that the proposed bill classified him as "unproductive" as well, since he was
unmarried. He had to do something. After a news report about Energy Unlimited's rising
profits and increasing workforce, he hatched a plan.

He knew from his prior observations that Camila and Jürgen were usually out on
Friday nights. Those observations were several months out of date, but he rented a car
and staked out the house in the hope their schedule hadn't changed. When they didn't
come home immediately after work, evidenced by the dark house, he knew he was right.

If he could find proof in the house that Camila, and probably Jürgen, was gay, he
could use that against her. He didn't think about how he could do that; humiliation drove
him.

He had no trouble finding shadowed access to the grounds. In the backyard, he
examined a glass door between the patio and the interior. He had brought some tools, but
they weren't ideal for breaking into a house. Banging the glass with a hammer didn't hurt
the glass, but made a lot of noise. He had more luck with a ten-inch screwdriver. Pounding
it into the crack between the door and frame opened a space, allowing the screwdriver to
penetrate deeper.

The tool was long enough to give him some leverage, and he bent the aluminum
frame significantly, but that was all. It was an old screwdriver, and the handle was worn
and rough; he stopped and tried to examine his throbbing hands, but it was too dark to
see if he was abrading his skin. He almost gave up, but he had already put so much effort
into the attempt. Swearing, he grasped the screwdriver again.

A loud pop startled him, and the door moved. Grinning, he pushed it open and
stepped into the house.

69

G REG WAS ONLY A few feet into the house before he collided with something. He could have brought a flashlight, but a nosy neighbor could see the light and know that the home should be empty. Somebody might have heard him force the door, and a sudden light would guarantee suspicion.

If there was any evidence that would prove Camila was gay, it would be in the bedrooms, probably upstairs. It seemed to take a long time to find the stairs, but a glance at his watch told him it was only four minutes. He was careful on the steps; this was no time to take a tumble. As he passed the landing halfway up, he saw a faint glow above, and that helped.

A nightlight was plugged into a socket in the hallway at the top of the stairs. That cast enough illumination for him to investigate the first bedroom and see a large room with a king bed and a bulky dresser. He couldn't make out more details than that. *Does Camila sleep with him in that bed?*

He shook his head and told himself this had to be Jürgen's bedroom. He moved farther down the hallway. Two doors down, he found what he was looking for, another bedroom that had to be Camila's. Again, the darkness prevented details, but there was a dressing table with enough bottles lined up on it to show that the bedroom was in use.

He wouldn't find anything else in the dark, but he was satisfied. He eased back to the stairs and down to the door he had forced. Outside, he was still grinning when a sudden flash of light blinded him, and somebody slammed him against a wall.

W HEN HIS PHONE BUZZED loudly, Jürgen put down a piece of pizza and pulled it out. "That's my silent alarm," he told the startled group. He looked at the phone screen. "Somebody is breaking into my house." He paused while the notice updated. "The police are already there." He could see fear in Camila's eyes.

She was trembling. "Our house? What should we do?"

"The police will handle it, but we should probably go home." He forced a smile. "More pizza for you," he told Mike.

"Should I come with you?" Mike asked.

Jürgen shook his head. "I don't see why. The police will be there when we get home. One of us will call when we have more information." He used his phone to summon a taxi.

An automated taxi was already waiting when Jürgen and Camila left the building. Their home was only five minutes away, and two police cars were parked in front of the house, their blinking, red lights sending strange shadows through the small crowd that gathered to watch.

An officer met them as they approached, and Jürgen identified himself.

"He broke in through the patio door," the policeman told them. "We caught him as he exited the house, but he wasn't carrying anything, so he wasn't stealing." He stared at Camila. "He was muttering something about looking for proof that you're a homosexual, but it didn't make a lot of sense."

Jürgen remembered the man that had confronted them outside La Parisienne. "Is he still here?"

The officer nodded. "We've got him in the back seat of that car. Do you think you know him?"

"Are you okay to look?" Jürgen asked Camila.

Camila only nodded slightly and walked toward the police car. Jürgen and the police officer followed.

As she approached, an angry face turned toward her and glared through the car window. "That's Greg Mora," she told the officer. "I dated him in college. I think he's been stalking me."

"What happens to him now?" Jürgen asked.

"He'll spend the night in jail and see a judge in the morning. I'll include your statement about him stalking you in my report, and the judge will see that, too. I don't think you have anything to worry about."

Jürgen nodded. *It won't be easy to convince Camila of that.*

P RESIDENT CORREIA STILL VALUED the friendship of Senator Valerie Peterson, and she was a frequent visitor to the Oval Office. His secretary brought her in, carrying two mugs of coffee. Peterson grinned and waved a reader.

"Good morning, Mr. President," she greeted as she approached the Resolute desk. "I brought you a present." She put the reader down in front of him as the secretary carefully placed the coffees on coasters.

Correia glanced at the reader with a frown. "I have a busy morning, Val. Can you summarize?"

"Sure. Somebody was caught breaking into Jürgen Varela's house Friday night in New York."

"The name sounds familiar. Why is he important?"

"He lives with Camila Lopez. Maybe you remember her better. She's the physicist that started Energy Unlimited."

Correia turned the reader to face him. "OK, yeah, I was at the ribbon-cutting for their factory. So what?"

"This guy told the police he needed to find proof that Lopez, and probably Varela, was gay. The police think he's a snubbed ex-boyfriend."

"Sounds like a soap opera plot. Why should I care?"

Peterson shrugged. "I thought you would find it interesting. Energy Unlimited has been very successful and is in partnership with a company in India. If Lopez is gay, there could be security implications around the technology."

Correia read from the reader for several seconds. "India is closer to Russia than we would like. It sounds like it's already a security problem if this technology is that important."

"I think it is. But so far, Energy Unlimited has protected their technology with secret patents. I don't know if Basu Power has access to details that would allow their products to be duplicated. Even if they don't, that could change."

Correia wanted to forget about the homosexual issue. His campaign pronouncements had helped him get elected, but that was done, and he had other concerns, both politically and domestically. Priscilla was not happy with him, and raising questions again would give her reason to doubt him more.

Correia pushed the reader back toward Peterson. "Considering homosexuality a security risk is bringing up some distasteful past."

"Lopez is apparently keeping it secret. And there's the possibility that Varela is gay too and uses her for cover."

Threatening to tax homosexuals was one thing; what Valerie was hinting at was something else. Still, she could be right.

"I'll have someone look into it," he promised.

70

T HE RESEARCH GROUP MET on most Monday mornings to update each other on their projects. "My design using the Stenhouse Field to direct the output flow instead of nozzles looks promising," Camila said. "I was worried about protecting surfaces outside the reaction chamber, but Rafael showed me data he'd collected. The field guiding the particles is more effective than we expected in protecting the surface."

Six years before, Rafael Rodriguez had helped Camila with her first prototype, using his work for his thesis toward a degree in Materials Engineering. After graduating, he joined Energy Unlimited.

"I saw his tests," Stenhouse said. "For some configurations, the field also protects the reaction chamber."

Ben leaned forward. "I should look at his results, too. Neutron damage will be an issue for my orbiting satellite design as well. It won't help my biggest problem, though."

"Controlling the output?" Camila asked.

"Right. The fission produces fast-moving particles that must go somewhere. That's a plus for a spacecraft engine, but a problem for an orbiting satellite."

"You were thinking about directing them tangentially so that they would cause spin rather than translation," Stenhouse said.

Ben nodded. "I'm still considering that, but we want to use these satellites to beam energy to receivers on Earth's surface. Spinning the satellite, and it would spin fast, would make that difficult."

"Two streams with opposite forces?" Camila suggested.

"Maybe. It seems a shame to waste all that energy, though. I can only extract so much from their motion in a small satellite."

"I have an idea." Stenhouse was smiling, a sight more common in recent months. "Design a toroidal recovery chamber where the field directs the particles in circles until most of their energy has been extracted."

Ben rubbed his chin. "That could work for the protons and metal ions. The neutrons would be more of a problem, though."

"Some materials can reflect them. Camila's suggestion of bringing in an engineer could help." Stenhouse scratched his head. "That makes me wonder if we could use the technology to build a particle accelerator."

The premier tools for particle physics research were still enormous facilities that used electromagnetic forces to accelerate particles to near-light speeds. The acceleration required distance, and circular or long paths directed the particles against a target with the desired energy.

Ben's eyes widened, and he smiled. "Possibly higher energies with cheaper facilities. That could revolutionize particle physics."

Stenhouse's smile faded a little. "I haven't gotten very far on the idea of a Stenhouse shield for spaceships. Maybe I should investigate this idea for now."

Disappointment made Camila hesitate, but she admitted to herself that this new idea would probably yield results in a shorter time. After discussing it some more, they agreed to suggest it at the next board meeting.

I N NOVEMBER, A POLAR vortex brought unusually cold temperatures and heavy snow to upstate New York. That didn't affect the scheduled board meeting at the end of the month, but Sahil and Amrita attended remotely rather than in person, as originally planned.

The meeting began on a positive note. Revenues continued to increase, and discussions had taken place with the Department of Energy about building an accelerator using Stenhouse technology. Basu Power had broken ground for Stenhouse nuclear waste treatment facilities at two of their nuclear power plants and the factory was manufacturing the reactors.

When Ryan opened the meeting for new business, the mood changed.

"The Russian Alliance and the Chinese have approached us about purchasing reactors," Sahil said from a screen on the conference room wall.

"Given the current antagonistic relations between the United States and those countries, I don't think that would be a good idea," Ryan replied.

"Our relations with those countries are much better. We could broker any agreements and assuage your government's concerns."

Ryan frowned. "I don't think the board would approve. While we are trusting you not to use the units we sell to you to reverse engineer the technology, we would not extend the same to Russia or China."

Sahil scowled. "Those countries have energy needs too. You know China is a significant source of the pollution flowing into your country. It would be fiscally irresponsible to sacrifice the potential profits and refuse the rest of the world for political reasons."

The Research Group planned short presentations on their work for the next day and Ben would talk about his power broadcasting satellite concept. Any implementation was probably well in the future, though.

"Fiscally perhaps, but our government is unlikely to let us ignore political and security reasons," Ryan said. "I suppose we could ask, but I think it's a nonstarter. I, for one, would be against it even if the government allowed it. Ben will, however, have a proposal later that could address the issue."

Sahil dropped the subject, but Ryan was certain it would come up again, in future discussions if not the current board meeting.

T HE MEETING RECESSED LATE in the afternoon, and Camila retrieved the jacket she had brought against the New York winter. Jürgen was talking to Ryan, but he would be ready to go home soon.

It had been three months since Greg's break-in, but it still bothered her. She supposed it was trite, but she felt violated and no longer safe in Jürgen's home. With the profits accruing from Energy Unlimited, she could afford a house of her own. But was that a solution?

She tried to forget her concerns as she approached Jürgen and Ryan. They were talking about Basu Power and the possibility of selling to Russia or China.

"I think Sahil knows he won't win this argument," Ryan said.

"We control the sales, but Basu has tied up a lot of their capital with their investment," Jürgen answered. "They could try to sell their interests to someone else to recover that capital."

"They could try, but our agreement with them stipulates that the board would have to approve of such a sale. We wouldn't do that."

Jürgen grinned. "I don't remember that clause being in the agreement I signed."

Ryan returned the grin and glanced at Camila. "You had references."

Jürgen also turned to Camila. "Ready to go home?"

"Sure." She wanted to delay, but didn't have a reason. Jürgen was studying her face, perhaps noticing her reluctance there. She forced a smile.

71

JÜRGEN NOTICED CAMILA CHANGE in the months after the break-in. He didn't think she noticed the increase in traffic on their street or had seen people stop and stare at the house, but they didn't talk about it, so he wasn't sure. Regardless, it still bothered her.

In the car returning from the office, she sat silently, a troubled look on her face. He had hoped her fear would fade, but it hadn't.

"Can we talk about it?" he asked.

Camila sighed. "Talk about what?"

"The break-in. It's obviously still worrying you."

She turned and stared at him. "Isn't it worrying you?"

"He's still serving the sentence for that. He won't be out for a couple months yet, and he'll have learned his lesson."

Camila shook her head. "Maybe. But he can still talk to people, telling them about us, and he's not alone. You were right to be afraid of the people who hate us."

"He was originally described as a scorned ex-boyfriend. I'm not sure anyone took him seriously."

"You haven't noticed the people who stop and stare at our home, trying to catch a glimpse of the 'gay people?'"

Jürgen sighed. "I have, but I was hoping you hadn't."

"I could afford my own place now. Maybe I should move out."

"How will that help? You moved here in the first place to make people think we're straight."

"They know where I live now. If I move, privacy laws will make it difficult to find me again and they'll forget about me."

She has a point. "Camila, you moved in here because we're friends and I didn't want people to know I was gay. In this climate, I'm never going to get married or have a relationship with another man, but I don't want to be alone either. I like having you around."

"I don't want to be alone either. But what can we do?"

"We could both move. We have more than enough money to move to a gated community that would provide security."

"Jürgen, I don't want you to uproot just for me."

Jürgen shook his head. "Why just for you? I have the same fears as you do. And I have a house, but without someone to share it with, it won't be a home."

"Are you sure?"

Jürgen grinned. "We're rich. We should think about something better anyway."

PRESIDENT CORREIA ROSE AS Spencer Sousa entered the Oval Office. "Spenser, welcome. It's been a couple of months."

His party chairman came forward and held out his hand. "Mr. President. Thank you for seeing me."

"Of course. Of course. Let's sit over there." He walked across the office, where couches provided a more comfortable place to talk. "What can I do for you?"

"As I'm sure you know, this is an election year. Midterms are always a problem for the party in the White House, and this year will be no different."

Correia nodded. He assumed the politician would suggest ways the president could help party members running for Congress, and Sousa didn't disappoint him.

"The party has appreciated your attempts to make nonproductive citizens pay their fair share," Sousa said. "There are concerns, however, about your apparent acceptance of climate change claims."

Correia raised an eyebrow. "I don't understand. I haven't made statements about climate change. It's a solved issue."

"Your support of that new energy company in your state gives a different impression. The company is quite vocal about the benefits of their 'clean' power, and I've gotten complaints from other companies in the industry. Especially the new fusion power plants."

"Energy Unlimited? I supported them because they created jobs in my state."

"I know, but that's not how their competitors see it, and they have more influence. There's another issue as well. Are you aware that the company is partly owned by Basu

Power in India? India is friendly with Russia and China. Our candidates think their opponents will claim that presents a security issue and will blame us for it."

Val had suggested that some officers of the company were homosexual and could be a security issue as well. "What do you suggest we do?" Correia asked.

B EN RAISED HIS BEER glass. "I want to make a toast," he told the other pizza eaters. "Two days ago, April 17, 2058, marked ten years since that momentous event that changed all our lives. Ten years since this woman," and he pointed to Camila, "asked me about the Stenhouse Field."

"Has it really been ten years?" Camila asked.

"It has. Two days ago exactly." He lifted his glass higher.

The others at the conference room table did the same, and several made comments or cheers. Ben lowered his glass and Gabby leaned over and kissed him on the cheek.

She smiled. "Changed indeed. I think we should mention something else, too."

Ben looked at his wife. "Now? You're ready?"

Gabby shrugged. "It will be obvious soon."

"OK then. Go ahead."

But Dani couldn't wait. "You're pregnant."

Gabby nodded and grinned. "The baby is due in November."

That triggered another toast, this time led by Ryan, with everyone asking questions at once. When there was a break in the conversation, Jürgen spoke. "Camila and I have another announcement. We're considering moving to a home in a gated community near Mariaville Lake. We decided we could afford something better now."

"And more secure?" Dani suggested.

Jürgen's expression was sober. "That too."

"Congratulations," Ryan said.

"It's about time we moved out of that apartment," Mike said. "Maybe we should buy a house."

Dani nodded. "We could look into the same community Jürgen and Camila are moving to."

"It's a fairly new development," Jürgen said. "There are homes available."

Ben looked at Gabby. She shrugged, and he took that as acceptance. "So let's hear more about this. We don't need to scrimp anymore either, and we're going to need more room."

72

R YAN'S SECRETARY ANNOUNCED A visitor. "She says her name is Chloe Winn." His secretary hesitated, and when he resumed, his voice was hushed. "I think she's a process server. Should I call your lawyer?"

Ryan frowned. *What's this about?* "Not yet. Send her in."

Chloe Winn was a nondescript woman about thirty years old. She looked nervous as she approached Ryan's desk and handed him a document. "You've been served, sir."

Ryan took the papers and glanced at the first page. "This is a subpoena for a Homeland Security inquiry."

"Yes, sir."

Chloe Winn was just a messenger. Questioning her for details would be futile. "Thank you, Ms. Winn."

She nodded and left the office. Ryan read the subpoena, but it only said that the inquiry concerned security issues with Energy Unlimited technology without going into detail. An addendum told him that Camila and Ben were also being summoned.

T HE MEETING WAS OFFICIALLY held in the Eisenhower Executive Office Building in Washington, but Ryan asked for and received permission to attend remotely from the conference room at Energy Unlimited. Ryan had a lawyer, William Peral, with him, in case they needed one.

Jaden Moore, a lawyer working for Homeland, opened the proceeding, explaining that concerns had been raised by President Correia about Energy Unlimited's relationship with India. He introduced the other four members representing Homeland and announced that they would begin by questioning Ryan.

Moore administered an oath and then began the questioning. "Mr. Ryan, you are the CEO of Energy Unlimited, yes?"

Ryan glanced at the others before answering. "I don't have that title officially, but yes, I have the usual authority for that position."

That began a brief discussion of EU's management hierarchy. They had never worried about titles; the core people had always reached consensus easily about who handled what. *Maybe that should change.*

"Perhaps the looseness of your management procedures is the cause of your problem," Moore finally said.

"I wasn't aware we had a problem," Ryan responded.

Moore gave him an annoyed glare. "That's what this inquiry is about, Mr. Terry." He paused and looked at the computer display to his left. Ryan couldn't see the screen, but Moore nodded as if he had found something. "Basu Power, an Indian company, owns part of Energy Unlimited, correct?"

Ryan nodded. "Twenty-five percent."

"And they are negotiating with Pakistan for Energy Unlimited."

"Yes. That was one of the terms in our agreement in return for the money we needed to develop our products."

Moore stared out of the screen. "And did that agreement also include the technology behind your products?"

Ryan kept his expression neutral. "No, sir, it did not. In fact, that has been the subject of some controversy with Basu Power. Our agreement stipulates that they would share in the profits and receive preferential access to our products, but would not receive our knowledge of the Stenhouse technology that makes our products superior."

"Couldn't they reverse engineer the units you sell them?"

"Not without a great deal of difficulty. They could, of course, disassemble a unit and get details of the reactor structure, but that wouldn't get much. It would take much more to decipher the design of the Stenhouse components, the control circuits, and the software that is at the heart of one of our reactors."

Moore didn't look convinced, but he moved on. During the next hour, Ryan answered questions from Moore and the other Homeland attendees about details of EU's business. A few questions had to do with the possibility of sales to other countries, and Ryan testified that there had been none yet and that Basu Power did not have the authority to broker such sales without permission from EU's Board of Directors. That started a discussion of the board's composition and assurances from Ryan that Basu Power would be outvoted if they proposed sales to other nations, especially Russia and China.

"I think that will be all for now, Mr. Terry," Moore finally said. "We would now like to question one of your scientists, Camila Lopez."

Camila also swore to tell the truth, and Moore asked his first question.

"Ms. Lopez, are you a homosexual?"

T HE SUDDEN QUESTION SHOCKED Camila into silence. She stared at Moore for a long moment.

Ryan reacted faster, leaning toward the screen with a clenched jaw and narrowed eyes. "That is out of line. The question violates her privacy."

"Miss Lopez is free to refuse to answer. And I will, of course, document her reluctance in my report."

The implication was obvious. Camila looked at Ryan for help, but he only glared at Moore. Camila wanted to see Jürgen, but he was busy elsewhere, arranging for their move to the new home.

Ben and Ryan looked at her questioningly. Moore stared at her too, but his face showed impatience.

"We're waiting, Ms. Lopez."

Ryan resumed glaring at Moore. "It's a stupid question. Just tell him."

It isn't a stupid question. They know what Greg said. She had lied to that reporter, but this was different. Ben took her hand, still looking confused. *Gabby hasn't told him.* If she told the truth, how would her friends react? This inquiry was about security concerns. In the past, people attempting to hide homosexuality had often been considered a security problem. Was her sexual orientation what this was all about?

Lying and being found out later would be worse. She swallowed and looked directly into the screen where Moore waited. "Yes, I am."

73

C AMILA'S QUESTIONING CONTINUED FOR several more minutes, but Moore had plainly gotten what he was looking for. Ben testified, but with only cursory questions about what he knew about Camila's sexual orientation and who else in Energy Unlimited might be homosexual.

"Camila's testimony came as a surprise to me. I don't know of any others either." Ben had to repeat that several times, but Moore dismissed him after ten minutes and turned back to Ryan.

"Mr. Terry, were you aware of Ms. Lopez's sexual orientation?" Moore asked Ryan.

"No. My knowledge in that area would be the same as Ben's."

Moore frowned. "I should remind you that you're still under oath. How long have you known Ms. Lopez?"

"About ten years."

"And you never suspected she was a lesbian."

"The thought has crossed my mind, but I never took it seriously."

"Can you explain why you considered the possibility?"

"Her lack of social life, I suppose. When I met her, she was dating someone, but that ended, and, to my knowledge, she dated no one else. Until Jürgen Varela, anyway. I didn't think about it after she met him."

"Mr. Varela is an investor in your company."

"Yes."

"And Mr. Pinto's brother-in-law."

"Also true. These relationships are part of the reason we haven't needed a fixed management hierarchy."

"But Varela is living with Ms. Lopez. Is he also a homosexual?"

"I have no idea."

"You don't seem to know much about your associates." Moore glared at Ryan.

Ryan returned the look. "Not about things that are none of my business. And shouldn't be yours."

"That's for us to decide, Mr. Terry."

W HEN THE HEARING ENDED, Ryan asked everyone to stay. "Who else knew, Camila?"

Camila hesitated. "Jürgen, of course. Dani and Gabby knew. I guess Ben didn't. I don't know if Dani told Mike."

"Probably not. If Mike knew, I think he would have told me."

"Why does it matter?" Camila asked. "It has nothing to do with Energy Unlimited."

Ryan's lawyer shook his head. "That should be true, but I don't think the government will see it that way. Keep in mind that this began as an inquiry into security concerns."

Camila hung her head briefly. "So what do we do?"

"Their primary concern was your relationship with Basu Power and India," Parel said.

"They own a piece of the company," Ryan replied. "Even if they would agree, we can't afford to buy them out. And we have a contract to supply reactor units to their facilities."

"You should wait for the government's next move," Parel said. "Your statements about protecting the technology may have satisfied them."

Ryan snorted. "Do you believe that?"

Parel grimaced. "Given Moore's attitude, it's not likely."

"Then I guess you and your firm have some work to do. We need to plan for their next move."

T HE NEXT DAY WAS Friday, and at first the talk at the usual pizza party was all about the previous day. Camila's revelation was a surprise, but not a shock, to most of the usual attendees. Jürgen admitted what everyone assumed anyway—that he also was gay.

He asked that the knowledge be kept within the group, and everyone was supportive of that.

Eventually, the conversation turned to Saturday's appointment at Mariaville Lake, when the three couples would complete their arrangements to move to their new homes. The meeting broke up soon after they agreed to meet at the Sales Office in the morning.

A T THE SALES OFFICE for the Adirondack Vistas gated community, they spent an hour signing final papers and arranging the delivery of their furniture and other belongings. The two salespeople who worked with them had questions about Energy Unlimited, adding time to the process. With the formalities complete, each couple went to their new home for a final inspection before the move.

Adirondack Vistas had a community center with a small coffee shop. When their inspections were complete, they met there and relaxed at a corner table.

"That was quite a surprise yesterday," Ben said. He looked at his wife. "Although I now understand a cryptic conversation between you and Camila years ago."

"He can be a little slow sometimes, but I love him," Gabby said.

"If I remember correctly, we had just gotten back from our honeymoon," Ben protested. "I had other things on my mind than Jürgen and Camila."

"I suppose you knew too," Mike said to Dani.

Dani smiled at Camila. "She never actually told me, but, yes, I knew."

"Doesn't your living together impede other relationships?" Mike asked.

"Theoretically," Jürgen answered. "For my part, I would prefer to avoid other relationships in the current environment. I can't speak for Camila."

"Both of us have had experiences that make us wary," Camila said. "Apart from our friendship, living together gives both of us cover from bigots." She said the last word with an angry edge.

"If Homeland releases a report on their inquiry, that's shot all to hell," Mike said.

Camila nodded. "I hope all this doesn't affect the company."

T

HE CABINET HAD ALREADY discussed several agenda items when President Correia recognized Homeland Security Secretary Mitchel Boehmer. "You have a report on security concerns with Energy Unlimited?"

"The full report should be on your readers," Boehmer said. "It includes the transcript of an inquiry held last week into security concerns over Energy Unlimited's connections to foreign governments."

"Specifically, India," Correia said.

"Yes sir. Also Pakistan. The company's representatives denied any ties to other nations. Of course, India is friendly with both Russia and China, so an indirect link is possible."

"Our concern is with the technology?" the Secretary of State asked.

"Primarily," Boehmer answered. "Energy Unlimited has used it only for power generation. Given the world's need for clean power, that's a significant issue, but the energy production involved is tremendous. The potential for a weapon is enormous."

"Doctor Stenhouse's work is public," the Department of Energy Secretary, Dylan Martin, said. "How do this company's products affect the potential for weapons?"

"They ship reactors to foreign companies," Correia said. "Could other nations learn more about exploiting this Stenhouse Field by taking the reactors apart?"

Boehmer shook his head. "According to the company, that's extremely unlikely. It's not Homeland's biggest concern."

Correia leaned forward. "And what is your concern?"

"At least one principal, the physicist Camila Lopez, is a homosexual. She admitted it to us but has kept her orientation secret otherwise. She lives with an investor in the company, Jürgen Varela, and we suspect he is also gay, and they live together to hide their homosexuality."

The Health and Human Services secretary frowned. "I'm surprised she would answer such an inappropriate question. And I suppose you consider her a security risk because of her sexual orientation."

Boehmer stared back at her. "I consider anything that a vulnerable person hides to be a security risk."

It wasn't the first time that the two Cabinet members clashed, and President Correia overrode any retort. "What are our options in dealing with this?"

"Energy Unlimited has contracts with India, including upgrading nuclear power plants to process waste," Boehmer said. "Basu Power owns a quarter of the company, but their CEO maintains the full board has control over sales to other nations. That doesn't prevent espionage."

"Using their secrets to extort knowledge of the technology," the president said.

"Yes sir. That is our concern. We also learned that one of their engineers is here on a visa and is still an Indian citizen."

Correia nodded. "We already have enough problems with homosexuals. All right, continue your investigation into the security dangers." He turned to the Department of Energy secretary. "I want your department to investigate the practicality of building a weapon. If the Stenhouse Field can be weaponized, we'll want ways to control information about it."

"What about our weapon development?" the Secretary of State asked.

Correia sighed. "A new weapons system, possibly more powerful than current nuclear bombs, is just what the world needs." He shook his head. "For now, let's just assess the possibility. Highest security level, of course. Keep aware of the risk that anything we learn could end up known by our enemies."

With that, the meeting moved on to the next agenda item.

BEN HAD NEWS AT the August board meeting. "My latest energy satellite design should resolve the rotation problem. Simulations on the ability to keep focused on a land-based receiver are promising."

Ryan smiled. "Congratulate Gabby for me. She and her people have done a fantastic job."

"How is Gabby?" Camila asked. "I passed her in the hall a couple days ago, but we haven't had time to talk. She's really showing now."

Ben beamed. "Her doctor says that the baby is healthy and a little larger than average for a mother just entering the third trimester." He looked at Ryan. "So when are you going to be a grandfather?"

Ryan shrugged. "Mike and Dani are waiting, I guess. Maybe moving into the larger house will make a difference, but I'm not pressuring them." He paused. "Are the satellites going to be a product soon?"

"I still have work to do on the design, but the biggest obstacle will be logistics. We have to launch the satellites or leave that up to the customer. Then there's a need to refuel them periodically. Even using Stenhouse fission, the initial fuel supply won't last forever. And we don't have a design for the receivers yet."

"Should we have more people working on it?" Ryan asked.

"Not yet, but maybe soon. You could get someone investigating the possibility of working with one of the space servicing companies."

Ben saw Camila frown and thought he knew what she was thinking. If she had her way, they would develop their own launch capability, too. But Energy Unlimited had a long way to go before it could deal with the issues that would entail.

Martina Patel was concentrating on the blueprint displayed on her workstation screen when Mary Welch opened the door to her office. Immediately, the small room filled with the clamor from the machine shop outside, but Mary closed the door quickly and handed Martina a phone.

Martina's own phone was in her purse in a locker elsewhere. She didn't get many calls, but those she got were seldom welcome, especially when she was working.

"It's Ryan." Mary handed the phone to Martina.

"I need you for a meeting in the EU offices," Ryan said.

"Now?"

"I'm afraid so. As soon as you can."

"What's this about?" Other than the Friday pizza fests, Martina rarely went to the office building. Her work was all in engineering support for Mary's machine shop.

"I'll tell you when you get here. We have a situation that involves you, but I don't want to say more over the phone."

That sounds a bit paranoid. It was highly unlikely that anyone would be listening in on the call. But after the Homeland inquiry four months before and how Camila had been bullied into admitting she was gay, anything was possible. "Okay, I should be there in half an hour."

The call ended, and she called a taxi. "I don't know if I'll be back today," she told Mary.

Martina's ride was waiting outside the shop. She arrived at the EU offices a few minutes later. "They're waiting for you in the conference room," the receptionist told her when she entered the building. The man didn't look happy, and Martina's stress level jumped up. Was this about her visa status? She had come to the United States originally to get her degree, and her permission to remain was extended when she joined Energy Unlimited. She lived knowing her situation could change almost instantly.

Ryan and Camila greeted her when she entered the conference room, but their expressions were even more somber than the receptionist's. She sat and waited with her hands tightly folded on her lap.

"Camila and I received subpoenas summoning us to a hearing in Washington in two weeks," Ryan said. "They had one for you also, and are probably at the machine shop by now trying to serve it."

"Homeland again?" she asked.

"Not exactly. The House Committee on Homeland Security." Ryan smiled, but it was weak. "We've hit the big time."

"Then it's not about my status?"

"Not directly. I think their concerns are with our relationship with India, though, so they summoned you."

Martina shook her head. "I don't know anything. I don't even work on EU projects anymore."

"We know that, but the government doesn't. I don't think you have anything to worry about."

"They could send me back to India."

Ryan nodded. "Maybe. I didn't think you intended to stay here forever anyway."

Martina had to think about that. "When I came here, I planned to return after I received my degree. All my family is in India." She sighed. "With all the excitement around Camila's work, I guess I got distracted."

She didn't say so, but her friends in Energy Unlimited were a second family, even though she worked at the machine shop and not the EU offices. Only the previous week, she had almost asked to be transferred to the factory where all the EU work was done now, but had put it off, not wanting to abandon Mary.

"I've asked my lawyer to join us and talk to us about the legal issues," Ryan said. "He's coming from Albany and should be here soon."

"Will this be a public hearing?" Camila asked.

"I don't know, but I doubt it matters. Any revelations will end up in the press, anyway."

Camila stared at her hands. Nothing had been made public about the Homeland inquiry, but a congressional hearing would not be discreet. Martina felt her friend's tension, but the realization that the investigation might be more about Camila than herself helped relieve her own stress.

Camila heard somebody talking to the receptionist.

"That must be Bill Parel," Ryan said. The lawyer came into the room a few seconds later, and Ryan briefed him.

"This is rather extraordinary," Parel said when Ryan finished. "It's hard to believe something like this would justify a hearing before a House committee."

"I'm afraid your firm is going to be busy putting together all the documentation they've demanded," Ryan said.

Parel nodded. "That's what you pay us for. And, given the brief time you've been in business, it won't be that hard."

Camila spoke up. "What are they going to ask about?"

"Judging from the earlier inquiry, their chief concern is national security." Parel rubbed his chin. "Summoning Ms. Patel would support that." He turned to Martina. "What is your role in the company?"

"I'm a mechanical engineer supporting our machine shop, Welch Precision Machining. I worked on the designs for our prototypes, but I haven't done much work for Energy Unlimited since the factory opened."

Parel nodded. "OK, I'm familiar with Welch, but not with the details of how they fit in."

"We bought them when we needed a facility to manufacture prototype components," Ryan said. "They had fallen on tough times after the owner's husband died. With our help, they are a successful business again, even without work from EU."

"I've never had access to the important technology," Martina said. "That's been fine with me, but I think it was also intentional." She hesitated and glanced at Ryan. "Is what I tell you covered by lawyer-client privilege?"

Parel shook his head. "Probably not. I represent officers of the company and Ryan specifically, but that wouldn't include you. It shouldn't be a problem, though, unless you're talking about something criminal."

"I don't think I've committed any crime. But an officer of Basu Power asked me to pass them information about EU." She shrugged. "But I don't have any information."

"When was this?" Ryan asked.

"About five years ago, when Amrita and Sahil came here for a board meeting. Amrita gave me a ride home and asked me to pass on anything I could. Maybe I should have told you."

"Amrita is one of the board members from Basu Power?" Parel asked.

Ryan frowned. "Yes. She's tried other ways to get information about our technology as well."

"She's a board member, so her attempts wouldn't be classified as corporate espionage." Parel hesitated and looked at Camila. "I have some advice you're not going to like."

Ryan waved a hand to tell Parel to continue. Camila turned away. Whatever the lawyer was about to say, she suspected she would be more upset than the others.

"Go on the offensive," Parel advised. "Anything that comes out in the hearing will be public anyway, so get it out there first and take the wind out of their sails."

Camila stared at a wall and murmured, "You mean me."

Parel nodded. "You and Ms. Patel. We should be able to book you on a talk show and get your story out before it gets twisted by some politician currying favor with supporters."

"I have to talk to Jürgen first."

C AMILA RAISED HER HEAD when the doorbell signaled a visitor, but Jürgen was already heading for the door. Except for others in their gated community, visitors would have required permission to come in, but the Pintos and Terrys would have called before dropping in.

"Are you expecting someone?" she asked, but Jürgen only grinned at her, and she took that as a yes. Curious, she followed him.

Jürgen opened the door and Camila was immediately embraced by her parents. Over her mother's shoulder, she saw Diego standing at the threshold with an uncertain smile.

Camilla pulled back far enough to stare at them. "Mama, Poppa. What are you doing here?"

Tomás scowled at her and shook a finger. "Did you think we wouldn't come to support our little girl? Jürgen told us everything, and we've taken a couple weeks' vacation to help you deal with these damnable politicians."

Nina looked worried. "Jürgen said you will be going on a network broadcast."

Camila nodded. "With one of my friends. Our lawyer advised us to get everything out before the hearing." She frowned. "I'm sorry. The program is a local D.C. broadcast, but the news will probably get to Texas. It could cause you some trouble."

Tomás shook his head. "Don't worry about that. I'll gladly tell off anyone who has a problem with you."

"We have a guest room for you," Jürgen told Camila's parents. "Diego, I'm afraid you're going to be sleeping on the couch, but it's a very nice couch."

"We could have gotten hotel rooms," Nina said. "We don't want to be a bother."

"No bother," Jürgen assured her. "Anyway, you can provide that support better from here than from some hotel."

Tomás beamed. "Then thank you very much for your hospitality."

C AMILA WAS NO LONGER the reticent woman who Kendra had introduced only a few minutes before. As she spoke, her tone hardened. She was incensed, and Kendra's anger grew with hers.

"I admitted my orientation to Homeland, so the malicious idea that it might make me vulnerable no longer applies, if it ever did."

"You're saying that the government already knows anything you might reveal at the hearing? Then why go to all this trouble?"

Camila shrugged forcefully enough to show her opinion even remotely. *I've got her wound up now.* But Kendra realized she was 'wound up' herself.

"I can't tell you what irrational reasons they have. I can only guess, and my guess would be politics. The politicians behind it are hoping to gain support for the midterm election."

Kendra's monitor showed the cameras were on her, and she nodded, sure that her viewers would see her agreement.

76

P RESIDENT CORREIA CONSIDERED CALLING a full cabinet meeting, but only needed two people, Homeland Secretary Boehmer and Energy Secretary Martin. They now sat across from him in the Oval Office.

"I assume you're aware of the Energy Unlimited interview yesterday on *Washington Views,*" Correia said. Both men nodded, looking unhappy.

"After that, is the hearing a good idea?" Correia's tone left no doubt about what the answer should be.

Boehmer shook his head. "They made it hard to call the company a security issue."

"It will look like we're singling them out because the one woman is a lesbian," Martin added. "If the House goes through with it, it will cost them votes and we could lose seats."

Boehmer sighed. "I was the one who asked the Homeland Committee to hold the hearing. I'll have them call it off."

"We still have a problem," the president said. "This could be the atom bomb all over again. With or without Energy Unlimited, other countries will work on exploiting the technology, especially for its potential for weapons."

"Given the energy involved, that is certainly an issue," Martin said.

Correia slapped his knee. "We have an advantage now. We must move forward before Russia or China do."

Martin frowned. "Develop our own Stenhouse weapons? That would start another arms race."

"We never ended the arms race," Boehmer said. "We don't publicize it much, but Russia kept it going when they deployed nuclear anti-satellite weapons. Much of our infrastructure is vulnerable to an attack in space that would limit our capability in a surface war."

Correia nodded. "Exactly. New, more powerful weapons could be the deterrence tool we need."

"What do you want us to do, Mr. President?" Martin asked.

"Cancel the hearing. We need cooperation from the company, so we'll send a delegation to them to discuss turning their research to our needs."

W HEN RYAN RECEIVED THE notification that the government had canceled the hearing but a delegation from Washington would visit Energy Unlimited, he called an emergency board meeting. Amrita and Sahil attended remotely, and Ryan included Martina since it affected her.

"We can't relax just because the House committee has backed off," Ryan told them. "General Herbert and Secretary Boehmer don't visit companies like ours for trivial reasons."

"Do they still have security concerns?" Camila asked.

"They're sending the Chairman of the Joint Chiefs of Staff and the Homeland Secretary. Whatever they want to talk about, it probably goes beyond ordinary security concerns."

"The military hasn't been involved up to now," Ben said. "There's only one reason for Herbert to come."

Ryan nodded. "We should have expected this. But are they worried that others might develop a Stenhouse bomb, or that they want to develop one?"

Ben shrugged. "Probably both. And they will want us to support them."

"That would involve the government participating in our business a lot more than it is now," Ryan said. On the conference screen, Sahil and Amrita were frowning and exchanging inaudible comments.

Sahil turned to look out of the screen. "How does this affect Basu Power?"

Ryan paused before he answered. "I don't know. It could increase concerns about your involvement in EU."

"Perhaps this is a good time to tell you we've been approached about our investment in Energy Unlimited, and we are considering an offer."

"You know the board won't approve selling your holding to Russia or China. Even if we did, our government would almost certainly step in."

Sahil nodded. "Yes, but the offer isn't from another country. It's from General Power. We are considering the generous terms and delayed bringing it to you until our management decided. This development will probably increase their desire to recover our investment."

General Power? They were a competitor, or had been, at least. As far as Ryan knew, they had ended their research into the Stenhouse Field after Energy Unlimited beat them to a product, and Stenhouse joined EU. But if India sold its interests to an American company, it would help in dealing with the government.

"I would oppose any weapons research," Ben said.

He probably was also thinking that Sahil's news would increase pressure for Energy Unlimited to work on projects with national security implications. Ryan wasn't comfortable with the idea either, but wasn't sure how much choice they would have.

Camila wasn't saying anything, but he could see how disturbed she was. She might side with Ben, but Jürgen would influence her, too. Jürgen would want to support Camila, but he might also be reluctant to oppose the government. Gabby might vote with Ben, but she wasn't timid about having her own opinion.

Basu Power probably would oppose military projects, but that would change if they sold to General Power. He guessed the American company would support the government.

If the issue came to a vote, it could divide the company into two camps. The result could rest with Camila and Jürgen, potentially creating a conflict in the Research Group.

The situation was getting interesting, and that wasn't a good thing.

"I have a proposal," Ben said.

77

THEY COULDN'T ATTEND THE meeting, but Camila's family had insisted on being nearby. Camila brought extra chairs into her office and let them wait there while she met with the delegation from the government.

"You and Martina did great on that show," Tomás said. "You'll be great with those politicians, too."

"I think so. Feel free to use my coffee machine or ask our receptionist if you need anything." She turned to her brother. "My computer has *A World of Your Own* installed."

Diego grinned. "You play games at work?"

Camila returned the smile. "With some additional physics, Jürgen's program is useful in doing simulations of our designs. You should remember that it can be used for more than killing monsters."

"Like going into bankruptcy?"

"Like practicing for the real thing." Camila stood. "General Herbert and Secretary Martin will be here any minute. I should go."

THE CONFERENCE ROOM WAS crowded. Ryan, Ben, Camila, and Sahil sat at one end of the long table as General Herbert and Energy Secretary Martin filed into the room and took seats. Ryan hoped he was hiding the mix of nervousness and awe as the President of the United States appeared on the screen on the opposite wall.

"We asked for this meeting to discuss our desire to explore weapons potential in your research," General Herbert said. "Can I assume that, as patriotic citizens, you will cooperate?"

"No." Ryan cursed inwardly. He hadn't meant to respond so abruptly, but it had come out that way. He saw shocked faces and tried to soften the bare statement.

"We believe we have something better to offer than weapons. Something that will provide more reliable deterrence for hostile actions from other countries."

General Herbert glared at Ryan and opened his mouth to speak, but Correia waved a hand and spoke first. "Peace without military superiority? Please, explain, Mr. Ryan."

"Not military superiority. Economic superiority, driven by control over the world's energy."

"Your power generation technology is a tremendous advance, but I think you might be overselling it," Energy Secretary Martin said. "You think selling your reactors to Russia or China will dissuade them from their ambitions?"

"We're not talking about selling them reactors," Ryan answered. "We want to sell them energy. What we are proposing will require substantial help from the government and will take years to complete, but the result will be energy independence for the entire world, controlled by the United States."

"Except for the influence India has because of their part ownership," Martin said. "And how are you going to accomplish this, anyway? Broadcasting power between continents? Nonsense."

"That could be part of it," Ryan admitted. "But only locally, not between continents. Our research group has been working on a different concept. A network of satellites would generate energy and broadcast to ground-based receivers. Receivers could distribute the power using existing electrical grids."

President Correia turned, probably looking at Martin on his own screen. "Is this practical?"

Martin shrugged. "Theoretically. Broadcasting microwave energy from space is possible. We're talking about a lot of satellites."

"We can make the preliminary research available to the government," Ben said. "But we've solved most of the problems, we think. Of course, Secretary Martin is correct in saying that you'll need a lot of satellites. We'll need government support for manufacturing and launching them."

"The key point is that the United States will control them," Ryan said. "Our analysis suggests that we could sell all the power any country needs for substantially less than the cost of their own facilities."

The scowl was gone from General Herbert's face. "If this can really be done, it could solve our biggest military vulnerability, Russia's anti-satellite weapons. If they depend on satellites for their power, using those weapons would hurt them as much as us."

"Couldn't they target our spy satellites and so on without destroying the energy satellites?" Secretary Martin asked.

Herbert shook his head. "I doubt it. Their anti-satellite capability is nuclear and not a precision weapon."

"And it would be possible to tell the satellites not to broadcast to their receivers," Ryan added.

"That all sounds wonderful," Herbert said. "But, with apologies to Mr. Nanda, what about the access India would still have and the potential security issues there?"

Sahil smiled. "Basu Power was proud to provide the funds for this worthy opportunity, but we would like to recoup our investment for other infrastructure projects, including the receivers Ben was talking about."

"General Power has already expressed an interest in purchasing Basu's interest," Ryan said. "There wasn't a security issue before, but eliminating any foreign ownership will remove the perception of an issue."

"I can see that this is worth pursuing," General Herbert said. "But I'm not entirely convinced it means we shouldn't research the weapons potential as well."

Ben shook his head. "In the 1940s, the United States developed the atomic bomb. Less than a decade later, Russia had stolen that secret from us. Would they have developed it if we hadn't first?"

Herbert stared at Ben. "Probably. The Nazis were working on it too, and Russia got some of their scientists after the war."

"Still, he has a point," Martin said.

"You want support from the federal government," the president said. "How much support are we talking about?"

"We don't have that worked out yet," Ryan answered. "Tens of billions, certainly. But it will mean less spending on alternative energy sources and create thousands of jobs."

General Herbert threw up his hands. "This is nothing more than a fantasy. I can't believe we're taking this seriously."

"We don't have to," Energy Secretary Martin said. "We could start with a government contract for continued research into the idea. It could be relatively small and determine whether this could be important or whether the general is right. The original reason for all this was concern over security risks. Mr. Terry has shown there is no risk, especially given Mr. Nanda's statement about selling their interests."

Ryan watched President Correia. The politician stared out at them with a slight frown, probably considering how to use what he had heard to benefit his party.

After a few seconds, Correia nodded. "All right then, let's move forward. We'll need legislation and to get that, we need a plan. Write a proposal to fund a study on the feasibility of what you have suggested. Secretary Martin will create a committee within his department to work with you and with the House Committee on Energy and Commerce to draft legislation." He stared at Ryan. "Is that acceptable?"

Ryan glanced at Camila and Ben before answering. "It is."

General Herbert was not satisfied. "What about weapons research?"

HIS SISTER HAD TEASED him about using *A World of Your Own* for killing monsters, and Diego admitted, at least to himself, that such fantasy scenarios had been his obsession for years. As Energy Unlimited grew into a successful company, making Camila rich, his outlook matured somewhat.

The company produced energy generators, but Camila had always dreamed of space travel. She had told them about the research she was doing, including possibilities that could someday make her dream a reality, and he became interested as well. He had been thinking about a new setting, different from the old monster scenarios he had grown up with. This was the perfect opportunity to explore his idea.

He sat at his sister's desk and admired her large workstation screen for a few seconds. "Play *A World of Your Own*," he commanded.

"Good morning, Diego. Shall I resume your current story?"

"No, let's start something new. The setting is a planet orbiting Alpha Centauri. Change some details randomly so that the planet is Earth-like, but different too. Reveal the details during play."

"The planet has been created. I have named it Camila."

"You named it after my sister?"

"This is her workstation. You can change it if you don't like it."

Diego shook his head. "That's fine. One thousand humans have settled the planet, and the goal is to maintain a thriving colony."

"Alpha Centauri is approximately four light years from Earth. Will Earth supply Camila?"

"Yes. They'll need a faster-than-light drive."

"Done. Earth and Camila have devices in orbit using Stenhouse Fields to provide a corridor of warped space that spaceships can use."

Diego chuckled. The Stenhouse Field was the technology behind Camila's company. She had said she used the program in her work, so it knew about the field. It was clever for it to use that knowledge to make up an FTL drive for his story.

He looked forward to telling Camila and Jürgen about the program's invention when their big meeting was over. He could tease his sister about how the program had thought of applications that hadn't occurred to her. It would have been even more exciting if the program had discovered something real.

But when the conference ended, the participants had much more to think about than Diego's game. The opportunity to mention it never arose.

Ben expected General Herbert to continue to push military applications. The possibility of their work being weaponized was something they should have discussed, but their plans for harnessing the Stenhouse Field for energy had dominated their priorities. He glanced at the others, but they remained silent, some frowning and some trying to ignore the general.

In his speech at the Nobel Prize awards, Hans Olsson had talked about resource scarcity as a cause of war, implying that the Stenhouse Field would help prevent conflict. Ben believed that, but General Herbert saw another application for their work. He couldn't stop people like Herbert from exploiting that, and maybe they were right. But he didn't have to be part of it.

Ben stood. "As head of the Energy Unlimited Research Group, I can say that we will refuse to work toward using Stenhouse technology as a weapon." He saw Camila nod.

President Correia glared from the screen. "You obviously are not speaking only for yourself." He sighed. "General, if you can convince us you need the weapons, we can consider investigating the possibility, but we can't force Energy Unlimited to do it. And they may be right about these satellites."

Ben sat down again, and Ryan nodded his approval. They had just won a battle, but the war was far from over. Camila looked happy. She probably saw the satellites as a step toward what she had wanted from the beginning, applying Stenhouse technology to space travel.

Her research on Stenhouse-powered spacecraft engines hadn't been a subject of the government meetings, but that research would also continue. He smiled at Camila, and she smiled back. She was excited about sending mankind to the stars, and that was something that excited him as well.

78

IT WAS THE TWELFTH week of the Pitcairn year, the middle of summer, sunny and dry, and with a hot wind doing little to temper the torrid temperature, even as the Early Night period approached. Goldstein would have liked to relax on his patio while he waited for Jean to finish reading, but knew it would be unpleasant. He settled onto a couch in the main room and watched his wife. She seemed engrossed as she read the last few pages of the book he had finished two days before.

Jean put her reader down on the table next to her and glared at him. "That's it? You've only told half the story."

"It's already too long." Goldstein rubbed his stomach. "What's for dinner?"

Jean gave him an exasperated frown. "If you wanted me to make dinner, you shouldn't have pressured me into finishing your book."

"I didn't pressure you. You seemed to like it, though."

"I did. How much of it is true, and how much did you add?"

"Mostly, I think it's true. Isaac has the historical facts and memoirs that Jürgen Varela and Diego Lopez wrote."

"Did Jürgen and Camila stay together? Did Gabby Terry have the baby?"

Goldstein grinned. "That will be for the sequel."

"You're going to write another? Finish the story?"

"Maybe. I'll have to eat something first, though. Since you spent the entire afternoon with your face in a book, I guess we'll have to call for a car and have dinner at Paulina's."

Jean nodded. "It's too hot for Pitcairn Chili, but a plate of their chicken salad and a glass of apple juice would work."

"It's always too hot for Pitcairn Chili."

"So let's get going. When we get back, you can start on the rest of Camila's story."

The End

At least until after dinner